MCGRAW-HILL TELECO...

Build Your Own

Trulove — *Build Your Own Wireless LAN* (with projects)

Crash Course

Louis	*Broadband Crash Course*
Vacca	*I-Mode Crash Course*
Louis	*M-Commerce Crash Course*
Shepard	*Telecom Convergence, 2/e*
Shepard	*Telecom Crash Course*
Bedell	*Wireless Crash Course*
Kikta/Fisher/Courtney	*Wireless Internet Crash Course*

Demystified

Harte/Levine/Kikta	*3G Wireless Demystified*
LaRocca	*802.11 Demystified*
Muller	*Bluetooth Demystified*
Evans	*CEBus Demystified*
Bayer	*Computer Telephony Demystified*
Hershey	*Cryptography Demystified*
Taylor	*DVD Demystified*
Bates	*GPRS Demystified*
Symes	*MPEG-4 Demystified*
Camarillo	*SIP Demystified*
Shepard	*SONET / SDH Demystified*
Topic	*Streaming Media Demystified*
Symes	*Video Compression Demystified*
Shepard	*Videoconferencing Demystified*
Bhola	*Wireless LANs Demystified*

Developer Guides

Vacca	*I-Mode Crash Course*
Guthery	*Mobile Application Development with SMS*
Richard	*Service and Device Discovery: Protocols and Programming*

Professional Telecom

Bates	*Broadband Telecom Handbook, 2/e*
Collins	*Carrier Grade Voice over IP*
Chernock	*Data Broadcasting*
Harte	*Delivering xDSL*
Held	*Deploying Optical Networking Components*
Minoli/Johnson/Minoli	*Ethernet-Based Metro Area Networks*
Benner	*Fibre Channel for SANs*
Bates	*GPRS*
Sulkin	*Implementing the IP-PBX*

Lee	*Lee's Essentials of Wireless*
Bates	*Optical Switching and Networking Handbook*
Wetteroth	*OSI Reference Model for Telecommunications*
Russell	*Signaling System #7, 4/e*
Minoli/Johnson/Minoli	*SONET-Based Metro Area Networks*
Nagar	*Telecom Service Rollouts*
Louis	*Telecommunications Internetworking*
Russell	*Telecommunications Protocols, 2/e*
Minoli	*Voice over MPLS*
Karim/Sarraf	*W-CDMA and cdma2000 for 3G Mobile Networks*
Bates	*Wireless Broadband Handbook*
Faigen	*Wireless Data for the Enterprise*

Reference

Muller	*Desktop Encyclopedia of Telecommunications, 3/e*
Botto	*Encyclopedia of Wireless Telecommunications*
Clayton	*McGraw-Hill Illustrated Telecom Dictionary, 3/e*
Radcom	*Telecom Protocol Finder*
Pecar	*Telecommunications Factbook, 2/e*
Russell	*Telecommunications Pocket Reference*
Kobb	*Wireless Spectrum Finder*
Smith	*Wireless Telecom FAQs*

Security

Nichols	*Wireless Security*

Telecom Engineering

Smith/Gervelis	*Cellular System Design and Optimization*
Rohde/Whitaker	*Communications Receivers, 3/e*
Sayre	*Complete Wireless Design*
OSA	*Fiber Optics Handbook*
Lee	*Mobile Cellular Telecommunications, 2/e*
Bates	*Optimizing Voice in ATM / IP Mobile Networks*
Roddy	*Satellite Communications, 3/e*
Simon	*Spread Spectrum Communications Handbook*
Snyder	*Wireless Telecommunications Networking with ANSI-41, 2/e*

BICSI

Network Design Basics for Cabling Professionals
Networking Technologies for Cabling Professionals
Residential Network Cabling
Telecommunications Cabling Installation

Cryptography Demystified

John E. Hershey

General Electric Global Research Center
Adjunct Faculty, Rensselaer Polytechnic Institute

McGraw-Hill
New York   Chicago   San Francisco   Lisbon
London   Madrid   Mexico City   Milan   New Delhi
San Juan   Seoul   Singapore   Sydney   Toronto

The McGraw·Hill Companies

1 2 3 4 5 6 7 8 9 0   DOC/DOC   0 8 7 6 5 4 3 2

ISBN 0-07-140638-7

The sponsoring editor for this book was Marjorie Spencer and the production
supervisor was Sherri Souffrance. It was set in Century Schoolbook by MacAllister
Publishing Services, LLC.

Printed and bound by RR Donnelley.

McGraw-Hill books are available at special quantity discounts to use as premiums
and sales promotions, or for use in corporate training programs. For more information,
please write to the Director of Special Sales, McGraw-Hill Professional, Two Penn
Plaza, New York, NY 10121-2298. Or contact your local bookstore.

This book is printed on recycled, acid-free paper containing a minimum of
50 percent recycled de-inked fiber.

Dedicated with love to:

Anna
  Jim
  John
  Dave & Kim
    Brooke
    Matthew
    Amber
    BreAnna
    Michael

CONTENTS

Contents

Introduction to Symmetric Cryptography

This first part comprises thirteen modules. These modules will familiarize you with some of the important terms and concepts of cryptography in general and symmetric of one-key cryptography in specific.

Symmetric cryptography is the classical genre of cryptography in which the sender and recipient both have, a priori, a common secret quantity. It is the use of this a priori secret quantity in developing a secure functioning cryptographic system, or cryptosystem, that is the focus of symmetric cryptography.

First Considerations

In this module, you will be introduced to cryptography in a rapid and unconventional manner. In studying cryptography it is necessary that you have the facility to appreciate, calculate, express, and compare very large numbers and very small numbers. It has been said that cryptographers deal with the largest and smallest numbers in the world. Two of the most important ways of doing this are with a mathematical function known as the factorial and with exponents. We will therefore jump right in and use the first of these, the factorial, in our discussion of the simplest of the classical cryptographic systems, the simple substitution cipher. We will also recognize two very important tenets of cryptography. These principles will become clear as we examine two simple, classical cryptographic systems.

Cryptography consists of many disciplines but is performed primarily for encryption and authentication. Encryption is the process of rendering information unreadable to anyone but the intended recipient(s). Authentication, as applied to persons or devices, seeks to ensure that a message has been sent by a party or device authorized to send it, or, a document has been created by the party represented as creating it. Authentication can also be applied to the contents of a message where encryption is used to ensure any alteration of the contents of a message by an unauthorized recipient can be detected.

The desire to protect information from interpretation or "being read" by an unintended party during transmission or other public exposure is an extremely ancient human quest. Though there are stories that go back further, one of the best known and good examples of this is the Caesar cipher. Supposedly, J. G. Caesar would encrypt some of his letters to Rome by using a simple case of what we now call a simple substitution cipher. In Caesar's cipher there was a *plaintext alphabet* (**P**) and a *ciphertext alphabet* (**C**). The alphabets were arranged as shown in Figure 1-1.

Plaintext is text presumed understandable to anyone who should view it. *Encryption* changes plaintext to *ciphertext*, text that is ideally not interpretable by any unauthorized person who should happen to see it. The *cryptoalgorithm* was to serially, letter by letter, replace each plaintext letter with the ciphertext letter directly below it. Thus, the plaintext DOG became the ciphertext GRJ.

Figure 1-1
A plaintext/ciphertext alphabet pair

P	A	B	C	D	E	F	G	H	I	J	K	L	M	N	O	P	Q	R	S	T	U	V	W	X	Y	Z
C	D	E	F	G	H	I	J	K	L	M	N	O	P	Q	R	S	T	U	V	W	X	Y	Z	A	B	C

Figure 1-2

A plaintext/ciphertext alphabet pair

P	A	B	C	D	E	F	G	H	I	J	K	L	M	N	O	P	Q	R	S	T	U	V	W	X	Y	Z
C	Q	R	Z	V	E	H	P	N	X	T	O	B	Y	C	G	I	U	L	A	J	M	W	D	F	S	K

The Caesar cipher eventually expanded into the simple substitution cipher in which the ciphertext alphabet is randomly selected, as in the example of Figure 1-2. The plaintext DOG encrypts to VGP in this example.

The Caesar cipher is more an encoding system as the ciphertext alphabet is not changed and once discerned is known. The general simple substitution cipher is a true cryptosystem, however, and like all cryptosystems it has a *keying variable* associated with it. The keying variable for a simple substitution cipher is the selection of the ciphertext alphabet.

We now have introduced in an elementary way the three ingredients one needs to know to operate a cryptosystem—namely, the cryptoalgorithm, the keying variable, and the plaintext for the sender or the ciphertext for the recipient. Theoretically, to decipher a secure system, one must know each of these three elements; although when evaluating a system, a standard assumption of professional cryptographers is that the adversary knows everything except the keying variable.

With all cryptoalgorithms, it is important to know how many keying variables there are. This is important because we will want to know the algorithm's resistance to "brute force" cryptanalysis. In a brute force approach, the adversary simply tries all possible keying variables until the decrypted plaintext is understandable.

How many keying variables then exist for the simple substitution cipher? We find the answer by the following argument. In how many ways may we select the letter corresponding to the plaintext letter A? The answer, for English, is 26. In how many ways may we select the letter corresponding to the plaintext letter B? If we said 26, we would be admitting the possible use of the letter we used corresponding to plaintext A. We would then have a utility problem because we could encrypt a message but we would be faced with an ambiguity in decryption, not knowing whether a plaintext A or plaintext B was meant. Therefore, we restrict our choice for the letter corresponding to plaintext letter B to any letter except the letter we picked to correspond to plaintext letter A. As a result, there are 25 choices for this second letter of the ciphertext alphabet, and the number of possible choice for the two letters is the product of 26 and 25. Continuing on in this fashion we find that the total number of possible ciphertext alphabets, and consequently the number of keying variables for the simple substitution cipher, is the sizable product

$$26 \times 25 \times 24 \cdots 3 \times 2 \times 1 \approx 4 \times 10^{26}$$

This product form is termed a *factorial* and denoted as 26!. In general, n-factorial is

$$n! = n \times (n - 1) \times (n - 2) \cdots 2 \times 1$$

The factorial is quite a useful function in cryptography and it behooves us to have an easily remembered and easily employed approximation to n-factorial for moderate and large values of n. This is provided by *Stirling's approximation*

$$n! \approx \sqrt{2\pi n} \left(\frac{n}{e}\right)^n$$

Substitution of a ciphertext alphabet letter or symbol is an example of general substitution, which is a classic genre of cryptography. Another of the classic genres is *transposition*. In transposition, the plaintext letters or symbols maintain their identity but their relative positions are changed. The keying variable for transposition is the specific method by which the relative positions are changed. As a simple example, consider that the plaintext is written into a frame or matrix and then extracted by a particular route. For example, let us assume that that our secret plaintext is

TOM IS AN AGENT

Let us ignore spaces and assume that the first step of a particular transposition process is to write the plaintext, left to right, top to bottom, into a 4×3 matrix as shown in Figure 1-3.

As the second part of our transposition example, let us assume that the route of extraction is to proceed up the columns right to left. The extracted letters are then

TGAMNASOENIT

Figure 1-3
The filled 4×3 matrix

T	O	M
I	S	A
N	A	G
E	N	T

Clearly the plaintext letters have stayed the same but their initial relative positions, which made them understandable, have been severely perturbed.

We have now prepared sufficiently to appreciate two tenets of cryptography. The first tenet is

> **The fact that a cryptoprinciple has a large number of keying variables does not, in itself, guarantee that the cryptoprinciple is a good choice for building a secure cryptographic system.**

An easy way to understand this tenet is to consider the simple substitution cipher. It is this type of cipher that makes its way into countless newspapers each day. The cryptosystem, as we have seen, has approximately 400,000,000,000,000,000,000,000,000 different possible keying variables. This is an enormous number and yet these puzzles are regularly solved. Certainly, then, they are not solved by trying each possible cryptovariable.

The second tenet is

> **The fact that a cryptoprinciple is complex does not, in itself, guarantee that the cryptoprinciple is a good choice for building a secure cryptographic system.**

To see this, consider that we attempt to make our transposition cipher example even more complex by doing a second transposition after the first, that is, we make our system a double transposition cipher. The keying variable we select for the second of the transpositions is the writing of the first transposed text into a 3×4 matrix as shown in Figure 1-4.

We extract by sequentially removing the letters top to bottom, left to right, and obtain

TNEGANASIMOT

Figure 1-4
The filled 3×4 matrix

T	G	A	M
N	A	S	O
E	N	I	T

We could, of course, have saved a lot of trouble if we had simply written the plaintext in reverse order! This also illustrates a maxim that multiple encryptions do not necessarily make a system more secure.

Exercise 1

The purpose of this exercise is to help you develop a critical eye and test your observational skills.

1. The following message was sent in World War I. It is an example of steganography, the hiding or concealment of a secret within a larger, innocuous looking host text. What is the secret message?

   PRESIDENT'S EMBARGO RULING SHOULD HAVE IMMEDIATE NOTICE. GRAVE SITUATION AFFECTING INTERNATIONAL LAW. STATEMENT FORESHADOWS RUIN OF MANY NEUTRALS. YELLOW JOURNALS UNIFYING NATIONAL EXCITEMENT IMMENSELY.

2. Edgar Allan Poe (1809-1849) is considered by many to be the "father of the modern short story." Poe had a remarkable mind and a wide range of interests. The following is one of his sonnets, "An Enigma." Study it and make an astute observation.

   "Seldom we find," says Solomon Don Dunce,
   "Half an idea in the profoundest sonnet.
   Through all the flimsy things we see at once
   As easily as through a Naples bonnet—
   Trash of all trash!—how can a lady don it?
   Yet heavier far than your Petrarchan stuff—
   Owl-downy nonsense that the faintest puff
   Twirls into trunk-paper the while you con it."
   And, veritably, Sol is right enough.
   The general tuckermanities are arrant
   Bubbles—ephemeral and so transparent—
   But this is, now—you may depend upon it—
   Stable, opaque, immortal—all by dint
   Of the dear names that lie concealed within't.

3. In England, during the socially troubled days of the mid-seventeenth century, a Royalist was imprisoned in Colcester Castle awaiting probable execution. A letter for him was delivered to his jailer and eventually made its way to his hands. He read it and soon thereafter made his escape from the chapel wherein he had been left alone to make his peace. This is what the letter said:

"Worthie Sir John: Hope, that is ye beste comfort of ye afflicted, cannot much, I fear me, help you now. That I would saye to you, is this only: if ever I may be able to requite that I do owe you, stand not upon asking me. 'Tis not much that I can do: but what I can do, bee ye verie sure I wille. I knowe that, if dethe comes, if ordinary men fear it, it frights not you, accounting it for a high honor, to have such a rewarde of your loyalty. Pray yet that you may be spared this soe bitter, cup. I fear not that you will grudge any sufferings; only if bie submission you can turn them away, 'tis the part of a wise man. Tell me, an if you can, to do for you anythinge that you wolde have done. The general goes back on Wednesday. Restinge your servant to command. — R.T."

Study the letter and see if you can find what the prisoner found.

4. If we were to use a simple substitution alphabet with 27 characters, A–Z, and a special character for a space, how many keying variables would exist for this system?

5. We will be getting into number theory later, so we should start to think about numbers and their characterization well ahead of time. The factorial representation that we encountered in this module is a good place to start.

   Consider the number 10,200. It ends with two contiguous zeros. With how many contiguous zeros does 200! end?

6. Continuing on with factorials, what is the largest value of n such that 9^n divides 250! without remainder?

Plaintext

The study of plaintext (the statistics, patterns, and other characteristics thereof) is one of the most important considerations of cryptography. It is important because it is absolutely essential for a cryptographer to know exactly what the cryptography is trying to conceal, which is the rich set of intertwined statistics exhibited by plaintext language. In this module, we examine two special classes of plaintext: written text and speech.

Natural language has redundancy. For example, if we encounter the letter Q, we almost invariably find that it is followed by the letter U. Natural language also exhibits significant frequency of occurrence differences between any size groupings of characters (n-tuples). The simplest, but extremely useful, set of such statistics is the one-tuple or monoalphabetic count—that is, the relative frequencies of A, B, C, ... X, Y, Z, and space. F. Reza[1] provides the expected monoalphabetic probabilities for a large sample of English text, which are shown in Table 2-1.

From the information in Table 2-1, we see that the space is the most common single symbol and E is the most commonly occurring English letter. Samuel Morse constructed his famous code based on a similar frequency of occurrence table. Using dots, dashes, and spaces for separators, Morse devised a code for English text that could enable a telegrapher to take less

Table 2-1

Monoalphabetic frequency probabilities for English text

Plaintext Symbol	Probability	Plaintext Symbol	Probability	Plaintext Symbol	Probability
A	0.0642	J	0.0008	S	0.0514
B	0.0127	K	0.0049	T	0.0796
C	0.0218	L	0.0321	U	0.0228
D	0.0317	M	0.0198	V	0.0083
E	0.1031	N	0.0574	W	0.0175
F	0.0208	O	0.0632	X	0.0013
G	0.0152	P	0.0152	Y	0.0164
H	0.0467	Q	0.0008	Z	0.0005
I	0.0575	R	0.0484	space	0.1859

[1] F. Reza, *An Introduction to Information Theory* (New York: McGraw-Hill, 1961), 184.

time to send a message than would be required if fixed-length code was used. The most commonly occurring letter, E, was represented by a single dot, the next most commonly occurring letter, T, was assigned a single dash, and a relatively rarely occurring letter, such as Q, was assigned a relatively long code (dash-dash-dot-dash for Q). Over a century later, David Huffman designed a code that could be used to significantly reduce the average length of a computer file by first studying the frequency of occurrence of basic file elements (for example, English letters) and then assigning binary code words whose lengths had an inverse relationship to the file element probabilities of occurrence.

Both the Morse and Huffman approaches squeeze, or remove, redundancy out of the plaintext. These are successful approaches because of the pronounced redundancy in English and other languages. In addition, this art, which is formally called *compression*, has continued to progress with the development and widespread deployment of more sophisticated algorithms.

These same frequency of occurrence statistics enable you to solve a simple substitution cipher. It is not a bad first guess to assume that for a simple substitution cipher the two most commonly occurring ciphertext symbols probably represent a plaintext space and plaintext E.

Speech requires separate consideration due to its unique problems and its rich, important, and costly cryptographic history. The human voice occupies a frequency spectrum that is well contained in the range of 0 to 4 kHz. Historically, a 4 kHz channel was available for many scenarios in which speech had to be transmitted. But how could you reasonably protect the transmission of speech? This question plagued communications engineers for multiple decades. In most cases, digitized speech, speech that was first converted into discrete symbols such as zeros and ones, was not an economical option, as we will see in the next section. As a result, many clever and truly ingenious schemes were devised for transmitting what came to be known as *scrambled speech* over a 4 kHz channel.

These schemes included things such as frequency inversion, band splitters, and *concomitant frequency translation* of the spectrum segments (analogous to the *substitution cipher*) and time segment rearrangement in which 4 kHz segments of speech were rearranged in relative time (analogous to a *transposition cipher*). The final development involved chopping up the speech into small segments of time bandwidth, followed by an apparent random reordering (*checker boarding*) of them for transmission. This subsequent scheme is analogous to a hybrid of substitution and transposition. The advantage of these different methods is that the scrambled speech does not require an increased bandwidth for transmission; that is, the scrambled speech could be successfully passed over a 4 kHz channel and descrambled at the receiver.

But speech has strong temporal and frequency interdependencies that are quite analogous to the redundancy of text, and what could be scrambled could often be unscrambled if enough effort was applied. The *sound spectrograph* is an important tool to use to do this.

The sound spectrograph is a very important technology for attacking or cryptanalyzing scrambled *analog* signals and, in particular, speech. It was developed by Mr. R.K. Potter and his colleagues during World War II. An extensive history of this technology and its capabilities is presented in a 1946 technical report.[2]

The sound spectrograph creates a display of the intensity of speech power in frequency versus time that is essentially a short-term power spectral density. As the NDRC report states, "It is the fact that both time and frequency variations are simultaneously displayed which makes spectrograms so valuable for decoding work."

Figure 2-1, which is taken from the NDRC book, displays a spectrogram of normal speech. Someone says, "one, two, three, four, five, six." Let's examine these words and try to get a feeling for the value of the spectrograph display. For convenience, we have inserted letters *a* through *k* at reference times.

Figure 2-1 displays two spectrograms of the same data. The upper spectrogram uses a band-pass filter (resolution) of 300 Hz, and the bottom

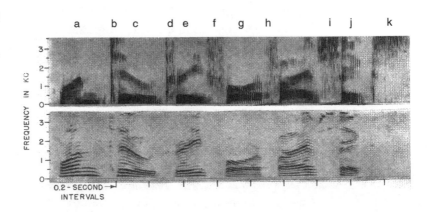

Figure 2-1
Spectrogram

[2] Summary Technical Report of Division 13, *National Defense Research Committee* (NDRC), vol. 3, *Speech and Facsimile Scrambling and Decoding*, Washington, D.C., 1946 (declassified, 2 August 1960).

spectrogram uses a much narrower bandwidth. The upper spectrogram resolution has been more widely used, so that is what we will discuss. Notice the dark wavy lines below a, c, e, g, and j, for example. These are called *formants*, which are the resonances of the vocal tract that are excited by the vibration of the vocal cords. This vibration is exclusively associated with *voiced* vowel sounds. Notice that with the spoken word "one," some of the formants rise with time, as shown below a. If you say the word, you can hear a rising frequency. Similarly, with the spoken word "two," you can hear a falling frequency, as shown below c. In contrast, the spoken word "four" seems fairly flat in pitch variation. This is disclosed by the formant picture in the spectrogram around time g.

Now let's look at the beginning of the spoken word "two" at time b. Notice that the spectrogram starts with a seemingly solid black line that indicates a lot of speech power at *all* of the frequencies displayed in the spectrogram. This is a different type of speech sound that is called a *plosive*. It results from a complete closure followed by a sudden opening of the vocal tract. No vocal cord excitation occurs, and the spectrogram discloses a short-duration wideband burst of energy. The same phenomenon is visible at time d at the start of the word "three."

A third class of speech sounds are called the *fricatives*. They are unvoiced sounds and have a sustained wideband spectrum as they sound like random noise. This is evident in sounds such as the f in "four" at time f, the f in "five" at time h, and the s in "six" at time i. The ending of "six" around time k contains an initial plosive followed by a fricative.

The only small (and portable) speech security device that was in factory production until the end of World War II used time-division scrambling. In this mode of scrambling, speech is chopped up into short segments and the segments are sent in what appears to be a random order. The spectrograph is used to analyze this unit's efficacy. It is even used to investigate two-dimensional scrambling, that is, the segmentation and rearrangement of the segments of a voice signal in both time and frequency. Figure 2-2, which was taken from the NDRC book, shows such a scramble in the upper spectrogram and its rearrangement in the lower spectrogram. Notice that the speech has been broken into many rectangular segments. The vertical extent of a segment is the frequency band of the segment; the horizontal extent is the time duration of the segment. The two-dimensional scrambling shuffles the relative position of the segments in an apparently random fashion. It is truly a jigsaw puzzle!

Undoing the time-frequency scrambling is very similar to solving a jigsaw puzzle. The beginning and end of the formants of one rectangle are matched with another rectangle, the pieces are put together, and so on.

Figure 2-2
Matching
spectrographs of a
two-dimensional
scramble

The critical point to note is that speech is a class of plaintext that has a lot of structure that can be exploited in cryptanalysis if it is left in its analog form—that is, if it is not first changed to discrete symbols. We cover this conversion to discrete symbols, called *digitization*, in the following module.

Exercise 2

The purpose of this exercise is to help familiarize you with the statistics and feel of plaintext.

1. What is unusual about the following sentence?

   JACKDAWS LOVE MY BIG SPHINX OF QUARTZ.

2. Ernest Vincent Wright authored a remarkable 267-page novel whose partial title is *Gadsby*. The Wetzel Publishing Company of Los Angeles published this book in 1939. The following is an excerpt from the book. What is remarkable about this passage (and probably about the entire book)?

   "Upon this basis I am going to show you how a bunch of bright young folks did find a champion; a man with boys and girls of his own; a man of so dominating and happy individuality that Youth is drawn to him as is a fly to a sugar bowl."

3. The monoalphabetic statistics of even small samples of randomly selected English text are usually quite close to the monoalphabetic frequency probabilities shown in Table 2-1. Do a frequency count on the following sentence taken from the *Electrical, Computer, and Systems Engineering* (ECSE) Department homepage of the Rensselaer Polytechnic Institute catalog for 1999 to 2000.

   "Electrical, computer, and systems engineers have been at the forefront of new discoveries and their integration into advanced design and engineering methodologies."

4. Solve the following cryptogram:

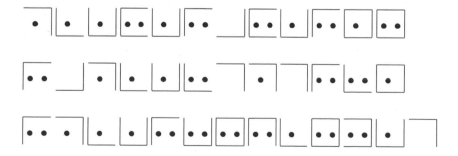

Figure 2-3
Spectographs from
question 5

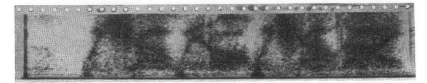

Specctrogram A

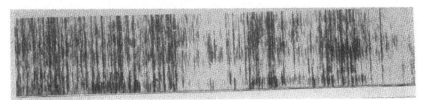

Specctrogram B

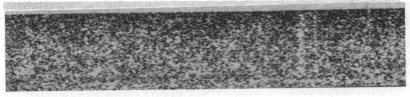

Specctrogram C

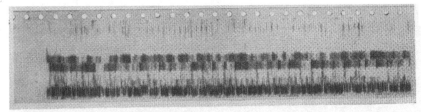

Specctrogram D

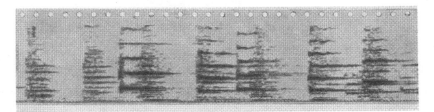

Specctrogram E

5. Figure 2-3 shows five spectrograms from the NDRC book. They contain
the following five sounds: a telephone bell, whispered speech, piano
music, crumpling paper, and random noise. Match them up.

6. A binary source is turned on. You observe that the first emitted 60 bits are (read bits left to right and top to bottom) as follows:

1 0 0 0 0 1 0 1 0 0 0 0 1 0 1 0 0 0 0 0 0 0 1 0 0 1 0 1 0 1

0 1 0 0 1 0 0 1 0 1 0 0 1 0 1 0 1 0 1 0 0 0 1 0 1 0 0 0 1 0

Based on the sample given, you conclude that the source exhibits redundancy. How many bits of information did the first 60 bits out of the source represent?

Digitization of Plaintext

Simply put, the modern world runs on bits; therefore, we need to adopt some way of representing plaintext in a digital form as a stream of bits. Many sources that need encryption are not inherently binary. As a result, we need to adopt a methodology to convert them to binary. This is not a particularly daunting task for text. The problem has been solved for many different situations in the frenetic history of modern telecommunications. This includes the use of a signal advance with the advent of the teletype and the Baudot code (which has five signaling bits per character framed by two synchronization elements). However, with voice, it has been a different matter. The analog-to-digital conversion of voice has followed a difficult and costly path primarily because of the large channel bandwidth that was initially required to support the transmission of good-quality digitized voice. In this module, we consider some of the basics of digitizing text and voice, and also look at the beginning of the secure voice effort in the United States during World War II.

For most examples in this book, we adopt the straightforward binary representation of text specified in Table 3-1.

Thus, DOG becomes 000110111000110. Note that our convention uses the binary encoding of Table 3-1 and converts the plaintext symbols from left to right.

But as we have said, the digitization of speech is far more complex. To gain an appreciation for the problem, let us first consider speech to be an

Table 3-1

A straightforward binary representation of plaintext

Plaintext	Binary	Plaintext	Binary	Plaintext	Binary
A	00000	J	01001	S	10010
B	00001	K	01010	T	10011
C	00010	L	01011	U	10100
D	00011	M	01100	V	10101
E	00100	N	01101	W	10110
F	00101	O	01110	X	10111
G	00110	P	01111	Y	11000
H	00111	Q	10000	Z	11001
I	01000	R	10001	space	11010

arbitrary and random time-based analog waveform called $s(t)$. This waveform may be considered the voltage that is delivered by a microphone serving a speaker. This and other components germane to our discussion are shown in Figure 3-1.

It is assumed that the analog speech waveform $s(t)$ does not contain significant spectral energy above 4 kHz, but it may be filtered first to ensure the removal of such energy. The waveform is then sampled to produce discrete time samples $s(n)$.

Sampling must be done according to the *Nyquist* criterion. This criterion specifies that in order for the sampled analog signal to be faithfully reproducible, the sampling must be at a rate that is at least twice the highest frequency present in the analog signal that is being digitized. Failure to comport to Nyquist's criterion results in a distortion known as *aliasing*.

The effect of aliasing and the reason for its name can be explained by a simple example. Figure 3-2 presents three cycles of a sine wave (top) that will be sampled. By Nyquist's criterion, we need at least two samples per cycle. The middle part of Figure 3-2 presents an oversampling using three

Figure 3-1
Speech digitization

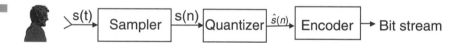

Figure 3-2
Analog-to-digital
sampling and analog
reconstruction

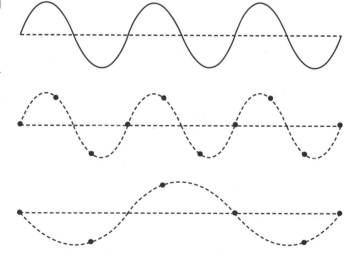

samples per cycle (denoted by large dots). The dotted line is the digital-to-analog reconstruction of the original waveform from the sample values. As you can see, the reconstruction is accurate. The bottom part of Figure 3-2 only uses 1.5 samples per cycle. The reconstruction resulting from this undersampling is highly distorted. The original sine wave has been reconstructed at half of the correct frequency. In other words, the original frequency has been aliased to a frequency that is half of the original. In Figure 3-1, we assume that the spectral content does not exceed 4 kHz; therefore, we sample at 8 kilosamples per second.

The samples are then *quantized*. Quantization associates a single level with all the values between two adjacent quantization levels. This introduces an error as the samples emerging from the *sampler* are assumed to be of infinite precision. The *quantizer* reduces their precision according to the number of quantization levels, which is 2^B. The encoder associates a B-bit binary word with each quantization level.

The quantizer can be chosen from a wide variety of instruments. For example, the quantization levels may be equally spaced, which is achieved by using a *uniform quantizer*. An example of this type of quantizer and

Figure 3-3
Uniform quantizer
with B = 3

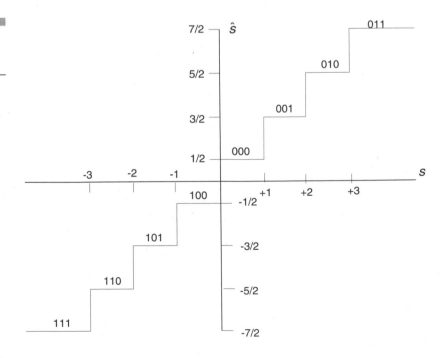

Figure 3-4
Sampling an analog
signal

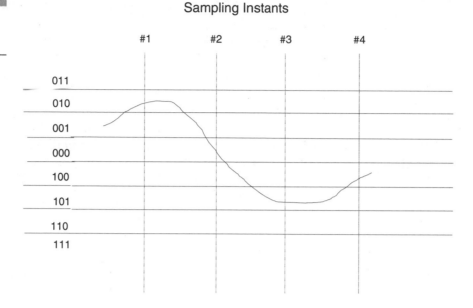

encoder (where $B = 3$) is shown in Figure 3-3. Note that if the input signal, S, lies between 1 and 2 volts, for example, the quantizer outputs the midrange value of 1.5 volts.

Figure 3-4 plots a segment of an analog signal, which might be voice, and shows how it might be digitized using the uniform quantizer and encoding of Figure 3-3.

The plaintext bit stream developed at the four successive sampling instants is

010000101100

Good voice quality can be maintained by using a uniform quantizer where $B = 11$, that is, $2^{11} = 2{,}048$ levels. Another type of quantizer spaces the quantization levels in an approximately logarithmic fashion. This keeps the quantization error approximately constant over the dynamic range of the quantizer, which is called *companding*. Examples of this type of quantizer are the μ-law and A-law logarithmic quantizers. It has been demonstrated that voice quality that is equivalent to that obtainable with a $B = 11$ uniform quantizer can be obtained by using a companding quantizer with $B = 7$. But here is the rub! If we need to sample at 8 kilosamples per second and encode each sample using 7 bits, we must convert our 4 kHz voice signal to a 56 Kbps binary stream, which does not fit into a 4 kHz channel.

Many complex techniques are currently available that provide good voice quality at a digital rate that can fit into a narrowband channel. It has been a long road of intense research and development. To help appreciate the problem further, the rich and fascinating history of the creation of a long-distance duplex secure voice link by the United States is worth considering. This development was motivated by World War II and the need for President Roosevelt and Prime Minister Churchill to be able to converse securely. In the article "The Start of the Digital Revolution: SIGSALY Secure Digital Voice Communications in World War II,"[1] Boone and Peterson set the stage:

> Before the full involvement of the United States in WWII, the United States and the United Kingdom were using transatlantic high-frequency radio for voice communications between senior leaders. The analog voice privacy system in use, called the "A-3," provided reasonable protection against the casual eavesdropper, but it was vulnerable to anyone with sophisticated unscrambling capability. This system continued to be used during the early part of the war, and government officials were warned that they could be overheard. In fact, it was later discovered that a German station in the Netherlands was breaking out the conversations in real time. This situation was intolerable, but neither the U.S. nor the U.K. had a ready solution.

The U.S. set about to build a secure digital voice communication system that came to be known as *SIGSALY*. SIGSALY was the first secure voice device to use quantization and the first useful instance of a speech compression technique. The analog-to-digital conversion of the analog plaintext was accomplished by a type of coder that segmented the voice signal into 10 frequency bands and measured the power within each of those bands. Six levels of quantization were used to represent the power in each band and the power was sampled every 20 milliseconds. Additionally, two channels were used to represent pitch information of the speech signal[2] as well as to indicate whether or not power resulting from unvoiced sound was present.

SIGSALY's voice quality wasn't great, but it was secure. This was attested to by a statement in 1943 by a previously successful German intel-

---

[1] "The Start of the Digital Revolution: SIGSALY Secure Digital Voice Communications in World War II," J.V. Boone and R.R. Peterson, www.nsa.gov/wwii/papers/start_of_digital_revolution.htm.

[2] Remember the formants we saw in the previous module?

ligence service regarding the interception of high-level Allied communications: "There is not much to be gotten from them now."

Although we have not yet spoken about combining a random digital stream (which is later known as *keytext*) with a digital plaintext, this was indeed done in SIGSALY. To do this, long stretches of keytext are produced and copies of the keytext are made on phonograph records for each end of the secure voice link. Boone and Peterson claim

> Key distribution, always a problem, was accomplished by means of transporting and distributing the phonograph records. These records were taken to Arlington Hall Station in Virginia, and the masters were destroyed. Once the systems were deployed, the key-pair was distributed by courier from Arlington Hall to the sending and receiving stations. Each recording provided only twelve minutes of key plus several other functional signals which were necessary for seamless key output.

> Later advances in recording technology permitted the simultaneous direct recording of the key on two acetate disks backed by aluminum. This technique reduced drastically the time required to make each record and also reduced cost. Since the keys were changed for every conference and on a regular schedule, there were a very large number of recordings made and distributed under strict controls. The key recordings were destroyed after use.

Keeping the keytext at the sending and receiving ends in synch was also a significant hurdle. Boone and Peterson explain

> . . . it was necessary for each record to be kept in synchronism within a few milliseconds for fairly long periods of time (one hour or so). This was accomplished by the use of very precisely driven turntables. The turntables themselves were remarkable machines. Each was driven by a large (about thirty pounds) synchronous electric motor with hundreds of poles. The motor was kept in constant operation, and the power for it was derived directly from dividing down the terminal's frequency standard. The frequency standard was a 100 kHz crystal oscillator. The accuracy of the standard had to be maintained within about one part in ten million so that the system would stay in synchronism for long periods of time. The system frequency standard could be corrected by comparing it to an available national frequency standard (which was WWV in the U.S.).

> Since one record held only about twelve minutes of key, it was necessary to have two transmit and two receive turntable subsystems at each terminal. In this way, a transition could be made from one key-pair to another.

All in all, the SIGSALY terminal comprised about 50 racks of equipment and weighed about 55 tons. Normal operation of the equipment lasted

Figure 3-5
A SIGSALY terminal

about 8 hours a day, and the remaining 16 hours were set aside for maintenance. Figure 3-5 is a picture of one of the terminals. Note the two sets of dual phonograph turntables on the right.

Exercise 3

The purpose of this exercise is to help you get a better grasp of the analog digitization process in regards to the approximation error.

1. Develop a simple model for the quantization noise developed within a quantizer step. Assume that the analog signal is uniform over the range of a quantizer step region.

2. Using the model developed in the previous problem, find the origin of the common wisdom that the signal-to-quantizing noise ratio is decreased by about 6 dBs per unitary increase in B.

Toward a Cryptographic Paradigm

Plaintext has a great depth of statistics that relate to the frequency of occurrence and linkage of its elements. Cryptography conceals those statistics and divulges as little as possible about the plaintext that exists behind the ciphertext. Now that we have agreed upon a technique for digitizing our plaintext, this section develops a paradigm for an important type of modern cryptography. We then turn our attention to the realization of that paradigm.

To proceed, we need to introduce some *Boolean algebra*. One of the most important of the two input Boolean functions is the *exclusive-or*, which is sometimes called the *half-add* or *addition modulo two*. It is defined by its truth table, which is shown in Table 4-1, where x_1 and x_2 are the two binary inputs and y is the exclusive-or of x_1 and x_2 written as $y = x_1 \oplus x_2$.

A number of properties of the exclusive-or are absolutely critical to our discussion. The first is that it is a linear operation. You can tell this because given *any* two members of the three-tuple $\{x_1, x_2, y\}$, it is possible to know or recover the third member. The second thing to note is that for any (binary) element $x, x \oplus x = 0$. In mathematical terms, we have an *Abelian group* of two elements.

A *group* is a combination of a set of elements $G = \{e_1, e_2, \dots\}$ and a binary operation $\circ$ so that the following properties obtain:

- For any two elements of G, $e_i, e_j \epsilon G$, $e_i \circ e_j \epsilon G$ (closure).
- For any three elements of G, $e_i, e_j, e_k \epsilon G$, $e_i \circ (e_j \circ e_k) = (e_i \circ e_j) \circ e_k$ (associativity).
- G possesses an operational identity element, $e_I \epsilon G$, such that for any element $e_i \epsilon G$, $e_i \circ e_I = e_i$.
- Every element of G, $e_i \epsilon G$, has an inverse, e_i^{-1} such that $e_i \circ e_i^{-1} = e_I$.

If, for any two elements of G, $e_i, e_j \epsilon G$, $e_i \circ e_j = e_j \circ e_i$, the group G is termed commutative or an Abelian group.

These options enable us to work with equations in a normal way so long as we keep to the properties of the exclusive-or. Thus, we can add the same

Table 4-1	x_1	x_2	$y = x_1 \oplus x_2$
The exclusive-or	0	0	0
	0	1	1
	1	0	1
	1	1	0

quantity to both sides of an equation and still maintain its integrity. So, if $y = x_1 \oplus x_2$, then adding x_1 to both sides gives

$$y \oplus x_1 = (x_1 \oplus x_2) \oplus x_1 = x_1 \oplus (x_2 \oplus x_1) = x_1 \oplus (x_1 \oplus x_2) = (x_1 \oplus x_1) \oplus x_2 = 0 \oplus x_2 = x_2$$

Now consider the construction shown in Figure 4-1.

Figure 4-1 depicts a module that generates keytext (a stream of ones and zeros) that is exclusive-ored bit by bit to a stream of plaintext to form a stream of ciphertext. One time instant is shown for which the ciphertext bit at time t (c_t) is formed from the plaintext and keytext bits at time t by $c_t = p_t \oplus k_t$. What can we learn about p_t from c_t? By using *Bayes' Theorem*, we have

$$\text{Prob}(p_t = 1 | c_t = 1) = \frac{\text{Prob}(c_t = 1 | p_t = 1)\text{Prob}(p_t = 1)}{\text{Prob}(c_t = 1)} \tag{4.1}$$

Now,

$$\text{Prob}(c_t = 1 | p_t = 1) = \text{Prob}(k_t = 0) \tag{4.2}$$

and

$$\text{Prob}(c_t = 1) = \text{Prob}(p_t = 1)\,\text{Prob}(k_t = 0) + \text{Prob}(p_t = 0)\,\text{Prob}(k_t = 1) \tag{4.3}$$

Therefore, Equation 4.1 can be written

$$\text{Prob}(p_t = 1 | c_t = 1) = \frac{\text{Prob}(k_t = 0)\,\text{Prob}(p_t = 1)}{\text{Prob}(p_t = 1)\,\text{Prob}(k_t = 0) + \text{Prob}(p_t = 0)\,\text{Prob}(k_t = 1)} \tag{4.4}$$

Figure 4-1
The conversion of plaintext to ciphertext

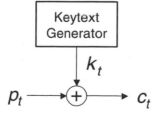

If k_t is known to be a zero with a probability of one-half (and therefore also a one with a probability of one-half), we cannot estimate p_t with any greater certainty by observing c_t under these assumptions. Equation 4.4 becomes

$$\text{Prob}(p_t = 1 | c_t = 1) = \text{Prob}(p_t = 1) \qquad (4.5)$$

It is also important that the keytext generator appear to have no memory; that is, the output at time t (k_t) must appear to be independent of what has preceded it. If this independence is lacking, then the preceding keytext bits (k_{t-1}, k_{t-2}, and so on) might help predict k_t. Our challenge during the next few modules is to see how we can approximate such a source of keytext bits.

The cryptographic paradigm is analogous to information theory. In information theory, we have a simple channel model called the *binary symmetric channel* (BSC), which is illustrated in Figure 4-2.

The BSC transports a bit correctly with the probability $1-p$ and inverts or transports the bit incorrectly with the probability p. The channel is memoryless in that the probability that it will cause an error at the t-th usage is independent of its past performance. The capacity of the channel (the amount of information it can transport per channel use) is denoted by $C(p)$ and is as follows:

$$C(p) = 1 + p\log_2 p + (1 - p)\log_2(1 - p) \qquad (4.6)$$

If $p = \frac{1}{2}$, then the BSC has a capacity of zero. In other words, if we are doing a good job of encryption, we are allowing the opposition to listen to our plaintext through a channel of zero capacity.

Figure 4-2
The BSC

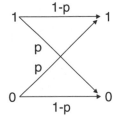

We desire the following two properties for the keytext:

- It must be balanced; that is, the probability of both a zero and a one must be one-half.

- It must appear memoryless; that is, the keytext bits before the keytext bit at time t must not aid in predicting the keytext bit at time t.

These two requirements specify what we term a *balanced binary Bernoulli source*. Such a source can be exemplified by the flip of a fair coin where the probability of turning up heads or tails is the same and the outcome is independent of previous outcomes.

A *general binary Bernoulli source* is still memoryless, but the probability that it will produce a one is p, which is not necessarily one-half; therefore, it is not balanced in general.

Two or more general binary Bernoulli sources can be combined to form a single binary Bernoulli source. Consider the situation depicted in Figure 4-3.

The probability that bit p_3, resulting from the exclusive-or of the two generalized binary Bernoulli sources, is a one can be easily found:

$$p_3 = p_1(1 - p_2) + p_2(1 - p_1) \tag{4.7}$$

Note (*and this is critical to our understanding*) that if one of the sources is a balanced binary Bernoulli source, the output (p_3) is indistinguishable from a balanced binary Bernoulli source. This is a simple application of Equations 4.1 through 4.4. Equations 4.1 through 4.4 go a bit further in that they show that any binary source combined with a balanced binary source is indistinguishable from a balanced binary Bernoulli source.

Figure 4-3

The combination of two general binary Bernoulli sources

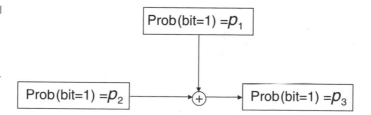

Prob(bit=1) $=p_1$

Prob(bit=1) $=p_2$ $\oplus$ Prob(bit=1) $=p_3$

Exercise 4

The purpose of this exercise is to help you understand some of the basic mathematical structures discussed in this module.

1. Is subtraction associative?

2. Is the multiplication of square matrices commutative? Investigate this for the specific case of 2×2 matrices.

3. There was once a computer architecture that enabled the contents of memory words to be operated on and the result of the operation placed into one of the memory locations; that is, imagine that there are two binary words, one at each memory location. An operation, f, on the words located in memory locations A and B, $f(A,B)$, produces a binary word that is placed into location B. What does the following series of operations result in if operator f is exclusive-oring bit by bit?

   $f(A,B)$
   $f(B,A)$
   $f(A,B)$

4. A couple has two children. You are told that at least one of them is a girl. What is the probability that both of the children are girls? (Try to work this problem using Bayes' Theorem.)

5. (The following is an example of von Neumann's coin.) You have a coin that is biased; that is, the probability of its coming down heads is $p \neq \dfrac{1}{2}$. In other words, the coin is an example of a generalized binary Bernoulli source. How can you use the coin to make a balanced binary Bernoulli source?

6. The following diagram shows five Bernoulli sources with different probabilities of producing a 1. They are exclusive-ored together. Find the probability that the output of the combined sources is a 1. Give your answer to four decimal places.

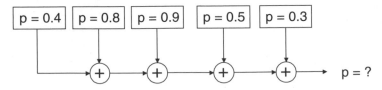

What We Want from the Keytext

In this module, we consider some of the properties that should describe the keytext. There are two extremely important considerations. The first is that the same keytext should never be used to encipher two different plaintexts. If this is done, a classic and serious cryptographic error occurs. It even has its own name—depth. The second consideration is that an inquiry should be made into the flatness of the keytext. Do ones and zeros have to be equally probable or can there be a slight bias?

The reuse of keytext is a serious error in cryptography. The system analyst must be acutely aware of the implications of this action and ensure that it does not happen under reasonable and foreseeable use.

The reuse of keytext is termed *depth* and two ciphertexts that have used the same keytext are said to be *in depth*.[1] Let's see why depth is so dangerous.

Suppose we have the following two ciphertexts:

CT1:   11010 11111 01010 00111 10100 00000 00001 10111 01101
       11000 01100 10101 00010 01000 00011 11000 11111
CT2:   01101 01110 00000 00101 01011 01100 11110 00001 11010
       01100 11100 10111 10111 10011 10111 01000 11100

Assume that we know that the plaintexts were converted per Table 3-1 of Module 3. Let $p_{i,j}$ represent the j-th bit of plaintext i, where $i = 1$ or 2. Similarly, let $c_{i,j}$ represent the j-th bit of ciphertext i and let $k_{i,j}$ represent the j-th bit of keytext i. Now, we have

$$c_{1,j} = p_{1,j} \oplus k_{1,j}$$

$$c_{2,j} = p_{2,j} \oplus k_{2,j}$$

However, if $k_{1,j} = k_{2,j}$ for all j, then we can get rid of the keytext as

$$c_{1,j} \oplus c_{2,j} = p_{1,j} \oplus p_{2,j}$$

for all j. Thus, the ciphertexts are linked to the plaintexts. We can use this to work one plaintext against another as we solve for both of them. The iterative procedure aims to develop one plaintext only so long as the other

---

[1]"The 1944 Bletchley Park Cryptographic Dictionary," www.codesandciphers.org.uk/documents/cryptdict/page27.htm.

plaintext makes sense. Let's start with the two previous ciphertexts: CT1 and CT2. First, we form $c_{1,j} \oplus c_{2,j}$ for all j. This gives us the following:

> CT1 $\oplus$ CT2:   10111 10001 01010 00010 11111 01100 11111 10110 10111
> 10100 10000 00010 10101 11011 10100 10000 00011

This is also, of course, PT1 $\oplus$ PT2.

We can start to work one plaintext against the other in many ways. One way to get started is to *drag* a *crib*.[2] A crib is a likely to occur plaintext word or phrase. Dragging the crib means trying it in successive positions. Well, that sounds eminently reasonable. Let's try the word FOR as the crib. FOR in binary is 00101 01110 10001 surrounded by spaces:

> 10111 10001 01010 00010 11111 01100 11111 10110 10111
> <u>00101 01110 10001</u>
> 10010 11111 11011
>   S ϕ^3 ϕ

This particular placement of the crib doesn't seem to yield potential plaintext. Let's try the next placement:

> 10111 10001 01010 00010 11111 01100 11111 10110 10111
>   <u>00101 01110 10001</u>
>   10100 00100 10011
>   U   E   T

The trigraph UET is not common, but it certainly occurs in English plaintext. Let's see if the development of plaintext holds up. To do this, let's check to see if the space following FOR yields something promising:

> 10111 10001 01010 00010 11111 01100 11111 10110 10111
>   <u>00101 01110 10001 11010</u>
>   10100 00100 10011 00101
>   U   E   T   F

Unfortunately, UETF no longer qualifies as developing potential plaintext so we move onto the next placement for our crib:

> 10111 10001 01010 00010 11111 01100 11111 10110 10111
>   <u>00101 01110 10001</u>
>   01111 01100 01110
>   P   M   O

---

[2] "The 1944 Bletchley Park Cryptographic Dictionary," op cit.

[3] We use the symbol ϕ to indicate that no plaintext character is defined for the 5 bits.

This placement also fails to develop potential plaintext so we try the next:

10111 10001 01010 00010 11111 01100 11111 10110 10111
<u>00101 01110 10001</u>
00111 10001 11101
   H   R ϕ

And try the next:

10111 10001 01010 00010 11111 01100 11111 10110 10111
<u>00101 01110 10001</u>
11010 00010 01110
  sp   C   O

Now this is beginning to look very promising. CO is a very common digraph for the beginning of an English word. Extending our crib FOR to include the bracketing spaces, we find the following:

10111 10001 01010 00010 11111 01100 11111 10110 10111
<u>11010 00101 01110 10001 11010</u>
11000 11010 00010 01110 01100
  Y   sp   C   O   M

This result continues our encouragement, and, as it turns out, we have posited a valid crib and its correct placement. We can now start working one plaintext against the other under the constraint that we develop reasonable plaintexts for both plaintext candidates. To reach the conclusion, we find that the two plaintexts are as follows:

TRY FOR TOUCHDOWN

and

EASY COME EASY GO

The lesson here (and it's a crucial lesson) is that the same keytext should not be used to protect two different plaintexts.

The second question for us in this module concerns the flatness of the keytext. Suppose it is slightly biased. It doesn't seem as though that would hurt us in general. How do we answer the question?

The answer is that we seek to provide a source for the keytext that is indistinguishable from a truly balanced, binary Bernoulli source because the user of the keytext should be able to use the keytext for any plaintext of interest and not have to even consider the question of balance.

As motivation, consider that many plaintext sources are highly redundant in that they repeat the same information many times. They are oversampled so to speak. One such source is associated with satellite telemetry. Consider the following remark taken from a paper by S. Rangarajan:[4]

> To control and effectively monitor the health of the satellite, telemetry data is essential. Onboard parameters are generally sampled much more often than the minimum Nyquist rate.
>
> · · ·
>
> The health of all the subsystems of the spacecraft is monitored continuously onboard and sent to the ground. Fast varying parameters are monitored in every frame whereas slowly varying parameters are monitored once in a master frame. Under normal circumstances many of the parameters are highly oversampled. In the case of geostationary satellites the thermal parameters, for example, generally vary with the period of one day, but are sampled once in a few seconds. Then there are parameters like the telemetry calibration voltages that are expected to remain constant throughout the mission life. These data, nevertheless, are to be monitored continuously in order to identify an anomalous condition which can suddenly develop on the spacecraft.
>
> · · ·
>
> There is, of course, a large amount of redundancy in the data. Very few parameters vary from frame to frame; and if they do (such as the attitude data), they vary so fast that there is hardly any necessity for knowing the specific value. However, as high as 84% of the bits in the master frame do not change under normal conditions. This is so because of unused words, used-up parameters (like deployment status), slowly varying parameters (like temperatures), constant values (like the most significant bits of onboard time), restricted operational range within the allotted wider range (like the wheel speeds), etc.

For our analysis, let's suppose that the keytext generator of Figure 4-1 in Module 4 was a binary Bernoulli source with a p (probability of a one) that was less than one-half without loss of generality. If we assume that the observed plaintext bit is equal to the observed ciphertext bit, then we will be correct with the probability $1-p > \frac{1}{2}$. In addition, if the plaintext bit does not change over n observations, we can have a much closer estimation.

---

[4]"Interpretation of Satellite Telemetry Data under Adverse Conditions," www.esoc.esa.de/external/mso/SpaceOps/2_17/2_17.htm.

Figure 5-1
Listening to satellite
telemetry

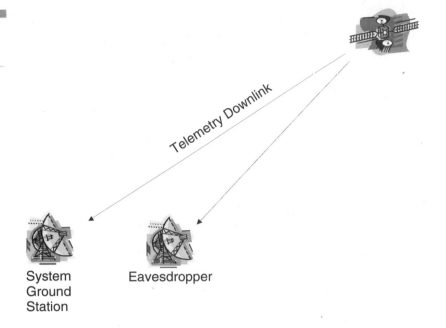

Telemetry Downlink

System
Ground
Station

Eavesdropper

We simply count the number of ones in the n-ciphertext bits. Let this count be denoted by c. Our decision rule is simply as follows:

$$\text{if } c \leq \left\lfloor \frac{n}{2} \right\rfloor, \text{ then assume the plaintext bit is a 0;}$$
$$\text{otherwise, assume it is a 1.} \tag{5.1}$$

The accuracy of our estimate is determined by the bias of the keytext ($\frac{1}{2} - p$) and the number (n) of ciphertext bits counted.

The distribution governing the statistics is the integral of the general binary Bernoulli source, which is the binomial distribution. According to the binomial distribution, the probability of there being exactly k ones in n observations of a binary Bernoulli source whose probability of producing a one is p is

$$\Pr ob(exactly \quad k \quad ones) = \binom{n}{k}p^k(1 - p)^{n-k} \tag{5.2}$$

where

$$\binom{n}{k} = \frac{n!}{k!(n - k)!} \tag{5.3}$$

The binomial distribution has the following mean and variance:

$$\mu = np$$
$$\sigma^2 = np(1 - p) \tag{5.4}$$

The probability that our decision rule will result in error (P_e) is

$$P_e = \sum_{k=\lfloor n/2 \rfloor + 1}^{n} \binom{n}{k} p^k (1 - p)^{n-k} \tag{5.5}$$

The form of Equation 5.5 is tedious to evaluate for a moderately large n. Fortunately, the binomial distribution can be well modeled by a normal probability density function for a moderate to large n when p is close to one-half.[5] To do this, we simply use the moments of the binomial shown in Equation 5.4 in the normal probability density function, and the probability of error shown in Equation 5.5 can be easily approximated by consulting a table of the normal distribution for the probability of an argument laying beyond the number of standard deviations (N_σ), where

$$N_\sigma = \frac{\frac{n}{2} - np}{\sqrt{np(1 - p)}} \tag{5.6}$$

We can turn Equation 5.6 on its head and solve for n in terms of N_σ and p, obtaining

$$n = \frac{4p(1 - p)}{(1 - 2p)^2} N_\sigma \tag{5.7}$$

---

[5]Mendenhall, Scheaffer, and Wackely in *Mathematical Statistics with Applications*, (Duxbury Press, 1981) advise that the normal approximation is "appropriate" for the binomial if $p \pm 2\sqrt{\frac{p(1 - p)}{n}}$ is in the interval (0,1).

Figure 5-2

n versus p

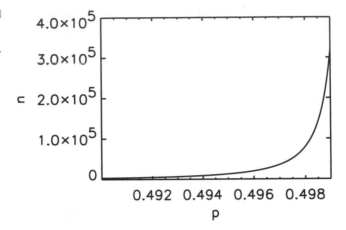

As an example, let's look at Equation 5.7 for a p that is in the range $0.49 \leq p \leq 0.499$. Let's say we want to be 90 percent certain that we make the correct decision. From a table of the cumulative density function of the normal, we find that we will be 90 percent correct if $N_\sigma = 1.2816$. In Figure 5-2, we plot n versus p for these conditions. Note that as p becomes closer to 0.5, the rate of increase of n tends to infinity as it must from Equation 5.7.

Another case in which it is desirable that the keytext be unbiased is that of traffic-flow security. The Department of Commerce/NTIA's Institute for Telecommunication Sciences in Boulder, Colorado, defines traffic-flow security as follows:

1. The protection resulting from features, inherent in some cryptoequipment, that conceal the presence of valid messages on a communications circuit; normally achieved by causing the circuit to appear busy at all times.

2. Measures used to conceal the presence of valid messages in an online cryptosystem or secure communications system. Note: Encryption of sending and receiving addresses and causing the circuit to appear busy at all times by sending dummy traffic are two methods of traffic-flow security. A more common method is to send a continuous encrypted signal, whether or not traffic is being transmitted.

Therefore, if we opt "to send a continuous encrypted signal, whether or not traffic is being transmitted," we will need to use keytext that is sufficiently well balanced or an eavesdropper may be able to discern when the circuit is busy and when it is idle.

Exercise 5

The purpose of this exercise is to bring home the critical concept of depth. The exercise will probably take you a number of hours. It may be tedious, but it is worth doing. You may want to write a simple computer program to help you test your assumptions as you unravel the two plaintexts.

1. It is believed that the following two ciphertexts are in depth. The plaintext is in the form of 5 bits per character as per Table 3-1 of Module 3. At least one of the messages contains the word THE. Solve for the *plaintexts*.

   CIPHERTEXT 1:
   **11001 10101 01110 01111 01011 00110 10001 00000
   10001 00011 10010 11011 00000 01100 01101**

   CIPHERTEXT 2:
   **01001 11001 10100 01110 10101 01111 00101 00011
   01111 01111 11001 11000 01101 10011 01001**

2. The following are the mod-2 bit-by-bit sums of two equal-length ciphertexts whose plaintexts consist of the characters A through Z and space, and were digitized per Table 3-1 of Module 3.

   **01011 00010 01110 10101 10000 00000 10010 10000 01011 11011 01001
   10010 10010 00011 00000 01111 10111 01101 00010 01000 00000 01001
   10001 10110 01111 11000 10101 11000 10101 00000 10101 11110**

   If the same keytext was used to encipher both plaintexts, how many times did the two plaintexts have the same character in the same position?

3. As we mentioned in Module 2, Huffman designed a coding technique that may significantly reduce the amount of bits needed to code plaintext that has symbols whose probability of occurrence varies over a wide range. Table 5-1 is the Huffman code for A through Z and space for English text.[6] Note that the most common symbols (for example, E, T, and A) have relatively short code words compared to those for the relatively rare symbols (for example, J, Q, and Z).

---

[6]F. Reza, *An Introduction to Information Theory* (New York: McGraw-Hill, 1961), 184.

Table 5-1

The Huffman code
for A through Z
and space in
English text

Symbol	Code Word	Symbol	Code Word	Symbol	Code Word
A	0100	J	0111001110	S	1100
B	0111111	K	01110010	T	0010
C	11111	L	01010	U	11110
D	01011	M	001101	V	0111000
E	101	N	1001	W	001110
F	001100	O	0110	X	0111001100
G	011101	P	011110	Y	001111
H	1110	Q	0111001101	Z	0111001111
I	1000	R	1101	space	000·

The Huffman code is known as a *prefix* code. This means that a Huffman code word of length k bits will never serve as the first k bits of a longer Huffman code word. Thus, the Huffman code is instantaneously decodable—that is, after the k-bit code word has been received, the decoder can immediately assign its symbol to it. The decoder does not have to wait for any additional bits to arrive before making its decision.

First, the coding method of Table 3-1 of Module 3 obviously requires 5 bits per symbol. How many bits per symbol does the Huffman code of Table 5-1 require?

Second, if plaintext was first coded by a Huffman code and then encrypted, why would it be much more difficult to solve for two plain-texts whose ciphertexts were in depth?

4. You know that a secret meeting is to be arranged and that the time of the meeting is going to be encrypted and sent. There are 12 possible times that the meeting may take place. They are all equally probable *a priori*—that is, before you observe or learn anything further. One of the following 12 messages was later encrypted and sent:

ONE_PM	TWO_PM	THREE_PM
FOUR_PM	FIVE_PM	SIX_PM
SEVEN_PM	EIGHT_PM	NINE_PM
TEN_PM	ELEVEN_PM	TWELVE_PM

The plaintext was encoded by the convention given in Table 3-1 of Module 3.

Assume that you observe the following ciphertext corresponding to the plaintext that has been encrypted.

010011011011000101010001010000100001

For each of the 12 possible messages, what is the probability that it was the one that was sent?

Randomness I

The production of good keytext is inextricably related to what we already know intuitively as a random process. But what does random mean, and once we decide that we need a source that appears random, how do we get it?

We have seen the need for a keytext generator to exhibit some very important properties. It needs to be evenly balanced in its output of zeros and ones, and appear to have no memory. In essence, we need a random or pseudorandom source that behaves as a balanced binary Bernoulli source of bits. Recall that a Bernoulli source is a memoryless source whose output at time t is independent of what it produced before time t. *Binary* means that its outputs are limited to zero and one; *balanced* means that the *a priori* probability of all outputs (in this case, zero or one) is equal. Thus, a balanced binary Bernoulli source outputs a zero at time t with a probability of one-half. A *pseudorandom* source's output is deterministic; a *random* source's output is not deterministic. A pseudorandom source appears random to an observer who does not know, or is unable to discern, the underlying determinism.

A fair roulette wheel is an example of a Bernoulli source. Typically, the wheel has three possible color outputs: red, black, and green (house takes). The *a priori* probability of a particular color showing up at the conclusion of a spin does not depend on the previous outcomes. This notion is inherently troubling to many people who viscerally feel that a monocolor output run should be less likely to continue at the next spin than a shorter run. Thus, in the early part of the twentieth century, when the roulette wheel at Monte Carlo came up red over 20 times in a row, the house collected substantially on heavy black betting near the twentieth occurrence of red.

On the other hand, if we present someone with the sequence

9265358979323846264338327950288841971693993751**x**

and ask for the next digit (the x), that person may decide that the sequence was randomly produced; therefore, the question has no unique answer. If, however, we preface the following sequence

3.14159265358979323846264338327950288841971693993751**x**

we see that we are being asked for the next digit of π, which is, of course, $x = 0$. In this case, most people fail to initially perceive the underlying determinism in that they do not perceive the initial string's likely relation to π.

Another case of failing to perceive the underlying determinism is afforded by the challenge to complete the following sequence:

OTTFFSSEN?

It is a consequence of our position as adults to have immense amounts of data stored in our minds for near-instant recall. Our thought patterns have matured, or at least changed, as we have aged and, many, if not most of us, would fail to perceive the sequence. Many, if not most, children in the higher elementary grades would perceive that the sequence is composed of the initial letters of the sequential positive cardinals; therefore, the next letter is T.

A third tenet of cryptography respects randomness:

Randomness is hard to get.

Let's try a simple analysis with a deck of cards. A deck of cards consists of 52 individually distinguishable elements. How many ways can a deck of these cards be arranged? If we stack them one on top of the next, there are 52 choices for the bottom card, 51 for the next, 50 for the one after that, and so on. The answer is 52!. By Stirling's approximation, this is about 8.05×10^{67}, which is a very large number. Now let's suppose we riffle shuffle a deck of cards. How many different single riffle shuffles are possible? The answer to this question may be easily bounded. First, the deck is cut in two at a randomly selected point and one of the two components of the deck is transferred to the left hand and the other to the right. Next, the cards in the two hands are melded or riffled together; that is, some cards from the left hand go to the deck being built, some from the right hand, some from the left, some from the right again, and so on until all 52 cards have been used to build the shuffled deck.

Assume that the deck is initially broken so that Λ of the 52 cards are placed in the left hand ($0 \leq \Lambda \leq 52$) and therefore $52 - \Lambda$ reside in the right. We ask for the number of ways in which we can select all 52 cards. Each possible selection sequence will be something like

$$L^{l_1}R^{r_1}L^{l_2}R^{r_2}L^{l_3}R^{r_3}\ldots \tag{6.1}$$

where $L^{l_1}R^{r_1}$ will be interpreted as l_1 cards selected from the left hand followed by r_1 cards selected from the right. The constraints are $l_i, r_i \geq 0$, and $\sum_i l_i = \Lambda$ and $\sum_i r_i = 52 - \Lambda$. The number of such sequences is $\binom{52}{\Lambda}$;

therefore, the number of such riffle shuffles is bounded by $\sum_{\Lambda=0}^{52}\binom{52}{\Lambda} = 2^{52}$.

Now, 2^{52} is a large number, but nowhere near as large as 52!. From algebra, we know that $a^x = b^{x \cdot log_b a}$, so $2^{52} \approx 10^{52 \cdot 0.3} \approx 10^{15.6}$. The consequence of this is profound because if we ask for the smallest integer (n) in order for $(10^{15.6})^n \geq 52!$, we find that $n = 5$; therefore, we have no hope of performing a random rearrangement of a deck of 52 cards without a minimum of 5 riffle shuffles. A random permutation appears *hard to get*.

Our goal, therefore, is to devise a source that is perceived to produce a random output stream of symbols to be used as the keytext.

As we have said before, a proposed pseudorandom source should be balanced in its output of symbols and appear memoryless. The latter requirement is the most challenging objective to fulfill. Checking for the appearance of memoryless behavior constitutes a rich statistical corpus. One of the most useful tools in this arsenal is simple correlation. As an introductory example, consider what is known as a *multiplicative congruential pseudorandom number generator*. This technique, incidentally, has formed the basis of Monte Carlo work and is the type of (pseudo)random number generator that is most often found in computer software packages.

For our example, consider the generator defined by

$$x_i = 3^i \text{mod}(19) \tag{6.2}$$

In practice, we develop the x_i's sequentially as a random number is called for. To do this, we start with an initial value or *seed*, x_0, with $1 \leq x_0 \leq 18$. We find x_1 by multiplying x_0 by 3, dividing by 19, and then keeping the remainder. We form x_{i+1} from x_i in the same manner. Thus, Equation 6.2 is made to function sequentially as

$$x_{i+1} = 3 \cdot x_i \text{ mod}(19) \tag{6.3}$$

where, again, the *mod* means that you divide by the argument in parenthesis (19) and keep the remainder of the division.

The result of successively replicating Equation 6.3 yields Table 6-1.

Our pseudorandom number generator appears to have a cycle length of 18; that is, it repeats after 18 steps. However, in any sequential 18 steps, it produces one of the possible 18 symbols; therefore, our generator is balanced as a pseudorandom generator that purports to generate integers in the range $1 \leq output \leq 18$. Now consider the graph shown in Figure 6-1 where we plot x_{i+1} versus x_i.

Table 6-1

The successive values produced by Equation 6.3

i	x_i	x_{i+1}	i	x_i	x_{i+1}	i	x_i	x_{i+1}
1	3	9	7	2	6	13	14	4
2	9	8	8	6	18	14	4	12
3	8	5	9	18	16	15	12	17
4	5	15	10	16	10	16	17	13
5	15	7	11	10	11	17	13	1
6	7	2	12	11	14	18	1	3

Figure 6-1
The graph of x_{i+1} versus x_i

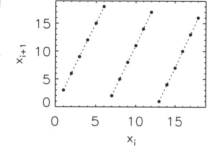

It is striking to see an apparent order emerge from what we thought might be a reasonable way to generate numbers with random properties. The advantage we have here is that we can take a multidimensional view, which, in this case, is a two-dimensional view. The results may be even more striking in higher dimensions. This was studied and captured in a descriptively named paper by George Marsaglia.[1]

As an example of something that requires a three-dimensional perspective to spot correlations, consider that we have a binary source that has put out the following stream of bits:

$$011101110000101000011110 \tag{6.4}$$

[1]George Marsaglia, "Random Numbers Fall Mainly in the Planes," *Proceedings of the National Academy of Sciences USA*, 61 (1968): 25–28.

Let's count the number of zeros and ones. We find 12 zeros and 12 ones. It seems well balanced if you look at it in one dimension. Let's count the di-bit frequencies. Table 6-2 shows the results.

The di-bits also seem to be well balanced. How about the tri-bits (the third dimension)? See Table 6-3.

Suddenly, with no hint or harbinger from the lower-dimensional analysis, we find a very definite and deep structure.

So, how much analysis do we need in order to safely pronounce something as sufficiently random? It's not an easy question to answer, but even though we do not answer it, we should constantly keep it in mind. There is perhaps limited solace in realizing that higher-dimensional analysis requires greater memory and processing time for both the cryptographer and cryptanalyst.

Table 6-2

Counting the di-bit or two-tuple frequencies

Di-bit Pattern	Number of Occurrences
00	3
01	3
10	3
11	3

Table 6-3

Counting the tri-bit occurrences

Tri-bit Pattern	Number of Occurrences
000	2
001	0
010	0
011	2
100	0
101	2
110	2
111	0

On a lighter note, another interesting and important aspect to memory-less binary sources concerns the expected waiting time to the selected sequence appearance. If we turn a balanced binary Bernoulli source on and observe the first three bits that come out of it, we would all agree that there are eight possibilities: 000, 001, 010, 011, 100, 101, 110, and 111. Each of the eight possibilities has an equal chance (one-eighth). However (and this is where the situation begins to become counterintuitive), if we asked a different question, such as "How long, on average, do we wait for a particular triple of bits to make its first appearance?" then the answer depends upon the particular triple. To ensure that we are on the same page, consider that we are looking for the first occurrence of the triple $B_1 B_2 B_3$ and that the source puts out the following bits after being switched on:

$$b_1 b_2 b_3 b_4 b_5 b_6 b_7 b_8 \ldots \tag{6.5}$$

Let's look at $b_1 b_2 b_3$. If this is the triple $B_1 B_2 B_3$, we have found it and only had to wait three time units. If it is not the triple $B_1 B_2 B_3$, we look next at $b_2 b_3 b_4$. If this is the triple $B_1 B_2 B_3$, we have found it and only had to wait four time units. If we continue to be unsuccessful, we keep on looking at $b_3 b_4 b_5$, then $b_4 b_5 b_6$, then $b_5 b_6 b_7$, and so on until we find the first occurrence of $B_1 B_2 B_3$.

A good way to get over the counterintuitive hump is to consider whether it makes sense that one triple might be more likely to occur first than another. Consider the two triples 000 and 100. The only way that the triple 000 can occur before the triple 100 is if the first three bits out of the source are all zero. The probability that the triple 100 will occur before the triple 000 is seven-eighths.

(Although we need not be concerned with it, the counterintuitivity gets far worse. For every particular triple, another triple is more likely to precede it. A good reference is provided by Martin Gardner.[2])

But let's get back to our original question. On the average, how long will we expect to wait before a given n-tuple appears for the first time from a switched-on balanced binary Bernoulli source? (The answer, and much

[2]Martin Gardner, "Mathematical Games," *Scientific American* 231, no. 4 (October 1974): 120–124.

more, is provided in a remarkable paper by Blom and Thorburn.[3] The calculation is quite simple. Let the n-tuple in question be the following:

$$B_1 B_2 \ldots B_n \qquad (6.6)$$

We first create a set of numbers, $\{e_r\}$, $r = 1, \ldots n$, where e_r is 1 if the first r bits of $B_1 B_2 \ldots B_n$ are the same as its last r bits—that is, if $B_1 B_2 \cdots B_r = B_{n-r+1} B_{n-r+2} \cdots B_n$. Otherwise, e_r is 0. We then form the following polynomial:

$$d(x) = \sum_{r=1}^{n} e_r x^r \qquad (6.7)$$

The mean waiting time for the first appearance of $B_1 B_2 \cdots B_n$ is $d(2)$, $d(x)$ evaluated at $x = 2$, and the variance is $d^2(2) + d(2) - 4d'(2)$, where $d'(2)$ is the derivative of Equation 6.7 with respect to x and subsequently evaluated at $x = 2$.

In the words of Donald Knuth,[4] "The moral of this story is that *random numbers should not be generated with a method chosen at random*. Some theory should be used."

---

[3]G. Blom and D. Thorburn, "How Many Random Digits Are Required Until Given Sequences Are Obtained?" *Journal of Applied Probability* 19 (1982): 518–531.

[4]Donald Knuth, *The Art of Computer Programming*, 2nd edition, (Reading, MA: Addison-Wesley, 1981), 5.

Exercise 6

The purpose of this exercise is primarily to help you gain a more visceral appreciation of the difficulty of achieving randomness.

1. We spent a little time looking at the mathematics associated with breaking and shuffling a deck of cards and we concluded that the number of permutations that breaking and shuffling could have on 52 distinct items was only about $10^{15.6}$, which is a large number, but much smaller than 52!. In this exercise, we're going to feel the implications of this disparity in numerical size.

   Arrange a deck of cards so that a basic sequence of suits is sequentially repeated through the deck. For example, the suit sequence CLUB, DIA-MOND, SPADE, and HEART (referred to as *CDSH*) might be chosen and the deck would be arranged as shown in Figure 6-2.

   Now introduce two elements of randomness into the ordering of the deck. The first thing to do is to build a pile of cards by taking cards off the top of the original deck one at a time. Continue this until the two

Figure 6-2
Arrangement of
the deck

H
S
D
C
.
.
.
H
S
D
C
H
S
D
C

piles (the pile being built and the pile supplying the cards taken one at a time) are of approximately equal size.

The second element of randomness is to riffle shuffle the two piles together.

Now examine the deck that has clearly been permuted by the two operations. If you take the cards off the top four at a time, you will discover that each group of four cards has one member of each suit. Your job now is to explain why this occurs.

2. You are using a data transportation protocol that is variable length and packet based. The structure of a packet is

FLAG	DATA	FLAG

where F is a FLAG field where the FLAG is 01111110. DATA is the data you want to send. You encrypt the data. Unfortunately, the encrypted data may have 8 bits in a row that are the same as the FLAG and unless this is somehow taken into consideration, the packet will be incorrect. On average, how many bits of encrypted data will go into the packet until a false FLAG is produced?

Finite State Sequential Machines

In this module, we look at some simple machines that are combinations of memory elements and Boolean logic gates. Some powerful concepts must be grasped, which can be gleaned from studying the behavior of these gates and memory elements. Modern cryptographic algorithms are based on combinations of these elements. We will need this information to examine the keytext generator in the paradigm of Figure 4-1 of Module 4, "Toward a Cryptographic Paradigm."

A set of memory elements serves as the heart of a finite state machine. For our purposes, these memory elements are binary and each element is therefore capable of being in one of two logical states. For convenience, we refer to these two states as one or zero. We diagram a simple memory element as shown in Figure 7-1.

The memory element has an input line that presents a bit (one or zero) to the memory element for storage, an output line that reports the state (one or zero) of the memory element, and a clock that commands the bit storage to be set to the bit value on the input line.

A common and useful machine is a linear array of 1-bit memory elements. Termed a *shift register*, it functions as a pipeline through which input bits pass. A four-stage shift register is shown in Figure 7-2. The clock is not shown in this figure. However, it is implicit that a common clock is supplied to all memory elements or shift register stages.

A more common representation of the four-stage shift register of Figure 7-2 is shown in Figure 7-3. Here we omit the interstage arrows of Figure 7-2 and it is understood that at each clock pulse the content of a particular stage is transferred to the stage immediately to its left with the exception

Figure 7-1
A solitary memory element

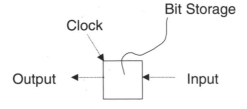

Figure 7-2
A four-stage shift register

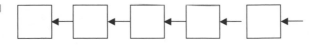

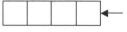

Figure 7-3
A more common way of representing the four-stage shift register

of the rightmost stage, which is assumed to accept its new bit from an explicitly drawn input line, and the leftmost stage, which, if left unconnected, does not pass its content on to another memory element.

The number of states that the four-stage register may assume is easily determined because each stage contains either a one or zero; therefore, there are 16 ($2\times2\times2\times2$) possible states. In general, an n-stage shift register has 2^n states.

We represent a state as a circle inscribed with the values of the memory elements as depicted in Figure 7-4.

The *state transition diagram* for a finite state sequential machine is the entire set of states. Each state is depicted as shown in Figure 7-4, connected by arrows that indicate the progression of states as the clock progresses. As an example, consider that we have a single memory element. We form two different machines as shown in Figure 7-5. The state transition diagrams are shown in Figure 7-6.

Figure 7-4
Representation of the state wherein the bits are a, b, c, d, respectively

abcd

Figure 7-5
Two machines formed with one memory element

This gate is an "inverter" - it turns a 0 to a 1 and a 1 to a 0

Figure 7-6

The state transition diagrams for the machines of Figure 7-5

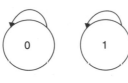

State Transition
Diagram for Figure
5 Left Machine

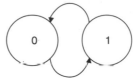

State Transition
Diagram for Figure
5 Right Machine

It is important that you now take some time to investigate the diverse behaviors of some simple and similar machines. Because of this, we are going to depart from the structure of the previous six modules and do the exercise in the middle of the module instead of at the end.

Exercise 7

Diagram the complete cycle structure for the following six 16-state sequential machines.

(a)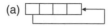

(b)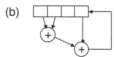

(c)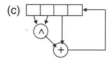

(d)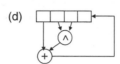

(e)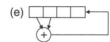

(f)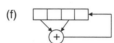

$\oplus$ Output = 1 if inputs are different; 0 otherwise.

$\wedge$ Output = 1 if both inputs are 1; 0 otherwise.

Transition Matrix

After having finished the exercise section, we now resume the module and have one more topic that must be covered before we move on. It is an important hierarchy of machine representation. The linear machine's *transition matrix* is at the first level of representation. If a machine has mod-2 feedback, such as Machines (a),[1] (b), (e), and (f), it is a linear machine. For a linear machine, we can write a matrix representation for the state transition progression. Let T designate the state transition matrix. For Machine (e), for example, we write

$$\begin{pmatrix} x_1^{t+1} \\ x_2^{t+1} \\ x_3^{t+1} \\ x_4^{t+1} \end{pmatrix} = T \begin{pmatrix} x_1^t \\ x_2^t \\ x_3^t \\ x_4^t \end{pmatrix} \tag{7.5}$$

---

[1]Machine (a) may be considered to have mod-2 feedback if we write $x_1^{t+1} = {}^=x_4^t \oplus 0$.

where

$$T = \begin{pmatrix} 0 & 0 & 1 & 1 \\ 1 & 0 & 0 & 0 \\ 0 & 1 & 0 & 0 \\ 0 & 0 & 1 & 0 \end{pmatrix} \tag{7.6}$$

where it is understood that the matrix computations are performed in mod-2. Note that we can find the state k steps away from any state by multiplying that state by T^k.

Also, because a machine such as Machine (e) is a linear machine, we can write

$$T \left[\begin{pmatrix} x_1^t \\ x_2^t \\ x_3^t \\ x_4^t \end{pmatrix} \oplus \begin{pmatrix} y_1^t \\ y_2^t \\ y_3^t \\ y_4^t \end{pmatrix} \right] = \begin{pmatrix} x_1^{t+1} \\ x_2^{t+1} \\ x_3^{t+1} \\ x_4^{t+1} \end{pmatrix} \oplus \begin{pmatrix} y_1^{t+1} \\ y_2^{t+1} \\ y_3^{t+1} \\ y_4^{t+1} \end{pmatrix} \tag{7.7}$$

where $\begin{pmatrix} x_1^t \\ x_2^t \\ x_3^t \\ x_4^t \end{pmatrix}$ represents one state at time t and $\begin{pmatrix} y_1^t \\ y_2^t \\ y_3^t \\ y_4^t \end{pmatrix}$ represents another

state at time t. For example, for Machine (e), the state 0100 is followed by the state 1001 and the state 0110 is followed by the state 1101. We would expect that the state 0010, the bit-by-bit mod-2 sum of the states x and y will be followed by the bit-by-bit mod-2 sum of the successors of x and y (0100), which it is.

The next level of the hierarchy is a machine that produces cycles but is not linear. Machine (d) represents this type of machine. You cannot write a relation such as Equation 7.5 for Machine (d). Machine (d)'s behavior may, however, be cast as a *permutation*. If we adopt the convention that the state number at time t, sn^t, is

$$sn^t = x_1^t + 2x_2^t + 4x_3^t + 8x_4^t \tag{7.8}$$

then we can describe Machine (d)'s transition behavior as a permutation:

$$\Pi_d = \begin{pmatrix} sn^t \\ sn^{t+1} \end{pmatrix} = \begin{pmatrix} 0 & 1 & 2 & 3 & 4 & 5 & 6 & 7 & 8 & 9 & 10 & 11 & 12 & 13 & 14 & 15 \\ 0 & 2 & 4 & 6 & 8 & 10 & 13 & 15 & 1 & 3 & 5 & 7 & 9 & 11 & 12 & 14 \end{pmatrix} \qquad (7.9)$$

The final level of the hierarchy consists of machines that do not produce cycles such as Machine (c). For these machines, the best that we can do is to write the state transition behavior as a *mapping*.

$$M_c = \begin{pmatrix} sn^t \\ sn^{t+1} \end{pmatrix} = \begin{pmatrix} 0 & 1 & 2 & 3 & 4 & 5 & 6 & 7 & 8 & 9 & 10 & 11 & 12 & 13 & 14 & 15 \\ 0 & 2 & 5 & 7 & 8 & 10 & 13 & 15 & 0 & 2 & 5 & 7 & 9 & 11 & 12 & 14 \end{pmatrix} \qquad (7.10)$$

m-sequences

The m-sequence is a result of finite field mathematics that has served in a variety of hardware systems. The sequence has excellent noise-like properties, is easy to generate, and can be cast in a linear framework. The latter property has enabled many parallel generation structures to be designed for the extremely high-speed generation of the m-sequence. The m-sequence may seem attractive to cryptography because of its remarkable statistical characteristics and its capability to produce a long cycle of states.

In Module 7, "Finite State Sequential Machines," we saw a profound difference in behavior between two very similar machines: Machine (e) and Machine (f). Both machines exhibit a single step cycle comprising the all-zero state, but Machine (e) links all the remaining states together in a single cycle with a length of 15 steps. Machine (e) is an example of what is known as a *maximal-length* or *m-sequence* generator. The term maximal length derives from the observation that a shift register with feedback built solely from addition mod-2 must exhibit a single step cycle if an n-stage shift register is set to, or started with, all zeros. The longest cycle that could theoretically be present is therefore one of $2^n - 1$ steps.

Let's take a closer look at the progression of the contents of any stage within an m-sequence generator. The following progression of bits is an m-sequence. If we start Machine (e) in the state 1000 and look at x_4^t, $t = 0, 1, 2, \ldots$, we see the following sequence:

$$100010011010111 \vee 1000 \tag{8.1}$$

In this case, the carat ($\vee$) indicates that the sequence starts to repeat at this point. Because the sequence repeats, 8.1 is said to be a particular *phase* of the sequence.

The statistics are what make an m-sequence so interesting. Let's count the frequency of occurrence of various k-tuples. First, we note that within its 15-bit period, the m-sequence of 8.1 has eight ones and seven zeros. This is close to balanced. Counting di-bits, tri-bits, and four-tuples, we arrive at the data in Table 8-1 and the statistics look quite attractive.

There is one other jewel. If we compute the cyclic autocorrelation of an m-sequence, it will be bivalued. This has a significant implication. To see why, let's first review what a cyclic autocorrelation is. We form the quantity R_k, which is the sum of agreements minus the sum of disagreements of an m-sequence displaced by k bits. For the sequence in 8.1, we compute R_l by writing

$$\begin{array}{l} 100010011010111 \\ \underline{000100110101111} \\ \text{daaddadaddddaaa} \end{array} \tag{8.2}$$

Table 8-1

The occurrence of di-bits, tri-bits, and four-tuples in a 15-step m-sequence cycle

Di-bit	Number	Tri-bit	Number	Four-tuple	Number	Four-tuple	Number
00	3	000	1	0000	0	1000	1
01	4	001	2	0001	1	1001	1
10	4	010	2	0010	1	1010	1
11	4	011	2	0011	1	1011	1
		100	2	0100	1	1100	1
		101	2	0101	1	1101	1
		110	2	0110	1	1110	1
		111	2	0111	1	1111	1

where a signifies agreement and d signifies disagreement. There are seven agreements and eight disagreements; therefore, $R_l = 7 - 8 = -1$. If we perform this computation for $k = 0, 1, 2, \ldots, 14$, we find that $R_0 = 15$ and $R_1 = R_2 = \ldots R_{14} = -1$. Thus, the cyclic autocorrelation has two values. The implication of this property is that it guarantees that the power spectral density of the m-sequence will be flat except for the DC value, the value for $k = 0$. It is this property, incidentally, that makes the m-sequence useful in direct-sequence spread spectrum system design.

An m-sequence exhibits a host of additional interesting and useful properties. The most celebrated of these is the *shift-and-add* property. This property proclaims that if an m-sequence is shifted by k bits, $k \neq 0$, and if bit-by-bit mod-2 added to its unshifted phase, it will produce the same m-sequence but at another phase. For example, let's add 8.1 to itself and shift it by three steps:

$$
\begin{array}{r}
100010011010111 \\
010011010111100 \\
\hline
110001001101011
\end{array}
\qquad (8.3)
$$

Note that the resulting sequence in 8.3 is indeed another phase of 8.1; it is 8.1 shifted by 14 bits.

Another property of an m-sequence is that it reproduces itself under *decimation* by a factor of two. The word decimation hardly seems fitting since it ostensibly implies something to do with 10, but the term has stuck. Decimation by a factor of k means that we take every k-th bit cyclically of a

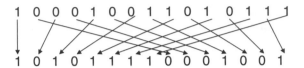

$2^n - 1$ step m-sequence until we have accumulated $2^n - 1$ bits from our decimated sampling. For example, let's use 8.1 again and decimate it by two as shown in Figure 8-1.

Note that the decimate-by-two sequence is just another phase of 8.1; this phase is reached by shifting 8.1 by 8 bits.

There is a phase of every m-sequence that reproduces itself if it is decimated by two. This is called the *characteristic phase*, and its existence and construction is taught by R. Gold.[1] For the m-sequence in 8.1, the characteristic phase is as follows:

$$000100110101111 \tag{8.4}$$

There is an m-sequence for every value of n and generally many for each value of n. The mod-two feedback connections for an m-sequence of length $2^n - 1$ are generally specified by what is termed a *primitive polynomial* of degree n. Such a polynomial is written as follows:

$$x^n + x^a + x^b + \cdots + x^c + \cdots + 1 \tag{8.5}$$

The convention is that the shift register's stages are numbered as we have done before, from right to left with an increasing stage number, beginning with stage 1, with bit flow through the register also right to left, and the feedback consisting of the mod-two sum of the contents of stages $n, a, b, c, \ldots$ For the m-sequence generator for 8.1, the polynomial is $x^4 + x^3 + 1$. This is realized by Machine (e)'s configuration.

---

[1]R. Gold, "Characteristic Linear Sequences and Their Coset Functions," *Journal SIAM* 14, no. 5 (1966): 980–85.

Table 8-2

A short table of primitive binary polynomials

n	Primitive Polynomial	n	Primitive Polynomial
2	$x^2 + x + 1$	12	$x^{12} + x^7 + x^4 + x^3 + 1$
3	$x^3 + x + 1$	13	$x^{13} + x^4 + x^3 + x + 1$
4	$x^4 + x + 1$	14	$x^{14} + x^{12} + x^{11} + x + 1$
5	$x^5 + x^2 + 1$	15	$x^{15} + x + 1$
6	$x^6 + x + 1$	16	$x^{16} + x^5 + x^3 + x^2 + 1$
7	$x^7 + x + 1$	17	$x^{17} + x^3 + 1$
8	$x^8 + x^6 + x^5 + x + 1$	18	$x^{18} + x^7 + 1$
9	$x^9 + x^4 + 1$	19	$x^{19} + x^6 + x^5 + x + 1$
10	$x^{10} + x^3 + 1$	20	$x^{20} + x^3 + 1$
11	$x^{11} + x^2 + 1$	21	$x^{21} + x^2 + 1$

There are many tables of primitive polynomials. W. Stahnke provides a primitive polynomial for n up through 168.[2] A short table constructed using Stahnke's data is provided in Table 8-2.

Remember that each $n > 2$ has more than one primitive polynomial. For example, $n = 3$ has two: $x^3 + x + 1$ and $x^3 + x^2 + 1$.

[2]W. Stahnke, "Primitive Binary Polynomials," *Mathematics of Computation* 27, no. 124 (1973): 977–980.

Exercise 8

The material comprising m-sequences is extremely rich. These few exercises touch on a few additional topics and, with luck, should engender some new thoughts.

1. The shift-and-add property of m-sequences provides a number of interesting and useful architectures and techniques. This problem is based on the m-sequence produced according to the primitive polynomial $x^3 + x + 1$. The generator is shown in Figure 8-2. What we will see in the solution of this problem is true in general and extends beyond this particular primitive polynomial of degree 3.

   The machine is started with register contents of 100. The cycle of the shift register contents is 7 bits long and is as shown in Figure 8-3.

   We attach a mod-2 summer to the machine of Figure 8-2. The summer has $n = 3$ inputs that are connected to the stages via switches S_1, S_2, and S_3, as shown in Figure 8-4. The mod-2 summer sums mod-2 the contents of the stages to which it is attached.

   Write down the 7 bits produced for each setting of the switches. Remember to start the machine in the original state of 100 each time. What does the machine in Figure 8-4 do?

2. We are using the m-sequence generator created according to the primitive polynomial $x^{10} + x^3 + 1$. A trial consists of initializing the shift register with 10 bits, running the shift register with mod-2 feedback for

Figure 8-2
The shift register m-sequence generator for $x^3 + x + 1$

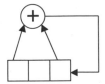

Figure 8-3
The cycle of states of the machine in Figure 8-2

S_3 S_2 S_1
1 0 0
0 0 1
0 1 1
1 1 1
1 1 0
1 0 1
0 1 0

Figure 8-4

m-sequence phase
generator

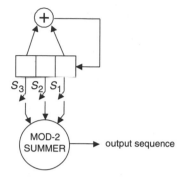

500 steps, and then reading out the 10 bits in the register after the 500 steps. Table 8-3 shows the results for four of these trials.

If we initialized the shift register to 1 0 1 1 0 0 0 1 0 1, what would the register contents be after 500 steps?

3. Suppose there was a lock that required a particular (non-all-zero) n-bit sequence to open it; for example, it might look like the device in Figure 8-5.

Imagine that the initial input register contents are zero. The lock opens as soon as the input register is filled with the secret combination. What is an efficient way to open this lock if you have no other knowledge of the secret combination?

Table 8-3

Pairs of register set-
tings displaced by
500 steps

Initial Register Setting	Register Contents After 500 Steps
0 1 0 0 1 1 1 0 1 0	1 0 0 1 0 0 0 0 0 1
0 1 1 1 0 1 0 0 1 0	1 1 0 1 0 0 0 1 1 1
1 0 1 1 0 1 0 1 1 0	1 1 0 0 1 0 1 0 0 0
1 0 0 0 1 0 1 1 0 1	1 0 1 1 0 1 1 0 1 1

Figure 8-5

The lock

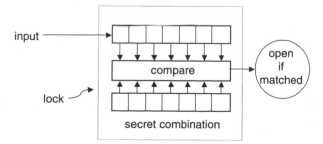

The Paradigm Attempted

It seems that m-sequences might take us well on our way to fulfilling the key-text generation paradigm of Figure 4-1 in Module 4, "Toward a Crypto-graphic Paradigm." m-sequences often have good statistical properties, are easy to construct, and could possibly be run at high speeds to accommodate modern cryptographic requirements. We will construct a cryptosystem based on an m-sequence and see if it actually provides all that is needed.

The m-sequence seems very promising as an approximation to a balanced binary Bernoulli source. How can we cobble it into a device that serves as a cryptosystem? First, let's review what we need from a keytext generator. We need the following:

■ Keytext that is a good approximation to a balanced binary Bernoulli source

■ Keytext that will not repeat for two different messages in order to avoid a condition of depth

However, there are other requirements; these requirements are addressed to the cryptographic system or *cryptosystem* that comprises the keytext generator. These requirements grow out of the possible use of a cryptosystem in a hostile environment. The first of these requirements is as follows:

The cryptosystem must provide security even under the assumption that the cryptosystem design is publicly known.

This is required because the security of a cryptosystem must not rest solely with its design, as eventually the design will become known through theft, loss, compromise, or analysis. The security of the cryptosystem must therefore also depend on something else.

This something else is known as the *key* or *keying variable*. It is a secret quantity that is inserted into a publicly known cryptosystem that configures or sets the cryptographic keytext generator in a unique state. This state is presumably unknowable to all parties except those possessing the secret keying variable.

Figure 9-1 illustrates a way in which we can possibly integrate our candidate m-sequence generator with a keying variable.

The keytext generator of Figure 9-1 functions as follows. Before encryption (or decryption), the keying variable is loaded into a storage register within the keytext generator. The keying variable is then transferred into the m-sequence shift register and provides the m-sequence generator with a secret starting point or secret phase.

It looks as though we might have a cryptosystem. But what happens if we want to encrypt more than one message? When we start to encrypt the

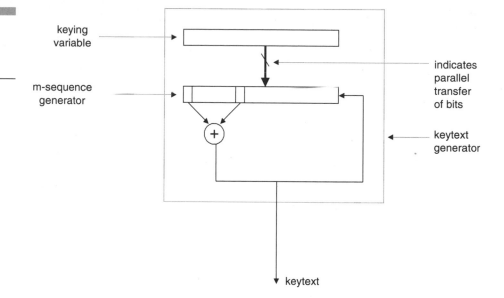

Figure 9-1
M-sequence
generator with
keying variable

keying
variable

indicates
parallel
transfer
of bits

m-sequence
generator

keytext
generator

+

keytext

second message, we will start the keytext generator at the same point in the m-sequence cycle that we used for the first message and the two ciphertext messages will be in depth. Clearly, we are not finished in our design.

This raises the following question: How do we provide different starting points to the m-sequence generator? One way is to add another cryptographic variable, but unlike the keying variable, this variable does not need to be kept secret. We'll call this additional variable the *initialization vector* (IV). An IV is randomly (or pseudorandomly) generated for each message and is either included as the preamble to a transmitted message or somehow associated with a particular message by a known and established convention. The device of Figure 9-1 is modified to that of Figure 9-2 to include the IV.

The keytext generator of Figure 9-2 functions as follows. The secret keying variable is loaded into a storage register within the keytext generator. When a message is encrypted (or decrypted), the IV for that message is entered into the keytext generator and added bit-by-bit mod-2 to the keying variable. Figure 9-2 shows a storage register for the IV, but the previous operation can be done in a variety of ways that do not involve storage of the IV within the keytext generator. The bit-by-bit mod-2 sum of the keying variable and the IV are transferred into the m-sequence generator's shift register and provide the m-sequence generator a secret starting point or phase that is unique to each message. As stated, the IV is communicated to

Figure 9-2
m-sequence generator with a keying variable and IV

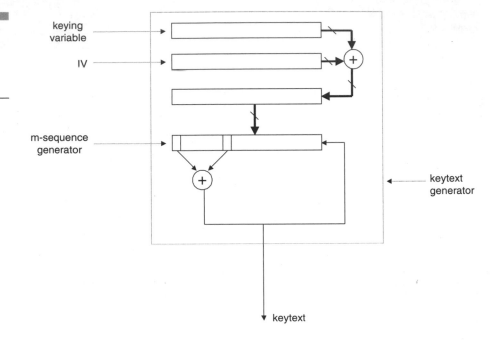

keying variable

IV

m-sequence generator

keytext generator

keytext

Figure 9-3
Bare-bones message format incorporating the IV and encrypted message that used that IV

| IV | ENCRYPTED MESSAGE |

the intended receiver openly by convention or with an encrypted message, as sketched in Figure 9-3.

Have we done it then? Have we discovered a viable candidate for the keytext generator paradigm of Figure 4-1 in Module 4?

Recall that when we studied depth we assumed that we could drag a crib through the ciphertext. The existence and correct placement of a crib means that we know a stretch of *matched plaintext* and *ciphertext*. It is not unusual to know matched plaintext and ciphertext. Various plaintexts have highly predictable statistics. Consider computer file data, for example. According to B.Y. Kavalerchik,[1]

---

[1]B.Y. Kavalerchik, "A New Model of Numerical Computer Data and Its Application for Construction of Minimum-Redundancy Codes," *IEEE Transactions on Information Theory* 39, no. 2 (1993): 389–97.

As a rule, numerical computer data redundancy exists in one of four forms:

1. The range of the data values is much smaller than can be represented with their storage format. This kind of redundancy is typical for business and commercial databases. For example, many numerical data items of customer order files, inventory files, etc. contain zeros or small integers preponderantly.

2. One or few data values occur with exceptionally high frequency. Mostly this value is zero.

3. Significant correlation exists between adjacent data values. This redundancy form is typical for sorted key values, telemetry applications, etc.

4. Integer values and values with a few significant digits in the fractional part occur considerably more frequently than values with all the significant digits in the fractional part. This kind of redundancy is typical for business and commercial applications. It occurs in the data rounded off to a given relative precision.

The four forms of redundancy overlap to some extent. For example, if the most frequent data value equals zero, then the second redundancy form is reduced to the first one. Storing the differences between successive data values reduces the third form to the first one, too. Therefore, the most conspicuous feature of numerical computer data is that they contain numerous leading and trailing zeros, with the latter usually in the fractional part. High frequency of the digits 0, as noted by many researchers, is a consequence of this feature. For example, in character frequency distribution for a typical commercial file given by J. Martin, the frequency of digit 0 is 16.3 times higher than the average frequency of any other digit.[2]

In telecommunications, it is not unusual to find instances where corresponding plaintexts and ciphertexts can be matched. A message classification system is available called *Encipher for Transmission Only* (EFTO). For this traffic, the content of a message is not classified.

As a general principle, therefore, a second requirement is added to what is expected of a cryptographic system:

The cryptosystem must provide security under the assumption that the opposition may amass as much matched plaintext and ciphertext as desired.

This second requirement can be stated in a number of equivalent ways. One way is to say that the prediction of the keytext bit at time $t + 1$ is not

---

[2]J. Martin, *Computer Data-Base Organization*, 2nd edition, (Englewood Cliffs, NJ: Prentice-Hall, Inc, 1977).

aided by knowledge of the keytext bits preceding it. (This is the memoryless aspect of a Bernoulli source.) Another way to say it is that the matched plaintext and ciphertext do not affect the security of the secret keying variable.

So what do these two requirements for the cryptosystem import regarding our candidate cryptosystem of Figure 9-2? Quite a lot. A $2^n - 1$ bit m-sequence is perfectly predictable if we have n consecutive bits. The requirement that the paradigm for the keytext generator must be a memoryless (Bernoulli) source is not met by such certainty. This is the principal downfall of our attempt to realize the paradigm with the system of Figure 9-2.

We have further to go, but we have developed a useful structure. The concept of an IV becomes essential as we move forward into modern cryptography.

Exercise 9

The purpose of this exercise is to further examine the interdependencies of the bits within an m-sequence.

The following three problems make use of Table 9-1, which is a duplicate of Table 3-1 from Module 3, "Digitization of Plaintext," for encoding the plaintext.

1. Assume that a cryptosystem uses an m-sequence to generate a keytext stream. Assume that you know that the three consecutive plaintext letters are AND and that the 15 ciphertext bits corresponding to the 15 plaintext bits are

$$1\,1\,1\,1\,1\,1\,0\,1\,0\,1\,0\,0\,0\,0\,1$$

Assume that the successive ciphertext bits are given from left to right and that the 5-bit plaintext letter coding is left to right. What is the minimum n that the shift register polynomial can be?

2. A five-stage shift register with feedback according to the primitive polynomial

$$x^5 + x^2 + 1$$

is used to produce a keytext stream, as shown in Figure 9-4.

Table 9-1

A straightforward binary representation of plaintext

Plaintext	Binary	Plaintext	Binary	Plaintext	Binary
A	00000	J	01001	S	10010
B	00001	K	01010	T	10011
C	00010	L	01011	U	10100
D	00011	M	01100	V	10101
E	00100	N	01101	W	10110
F	00101	O	01110	X	10111
G	00110	P	01111	Y	11000
H	00111	Q	10000	Z	11001
I	01000	R	10001	space	11001

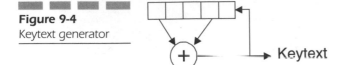

Figure 9-4
Keytext generator

The first 10 bits of ciphertext (left to right and corresponding to the 5-bit plaintext encodings that are also left to right) are

$$0011000100$$

Find the one digraph (two-letter pair) in the following set of five digraphs that is a viable candidate for the corresponding pair of plaintext letters:

SO   ME   DR   LA   AN

3. Assume that a cryptosystem uses a five-stage shift register with linear feedback to generate a keytext stream. Assume that you know that two consecutive plaintext letters are TH and that the 10 ciphertext bits corresponding to the 10 plaintext bits are

$$0\ 0\ 1\ 1\ 0\ 0\ 0\ 0\ 1\ 1$$

Assume that the successive ciphertext bits are given from left to right and that the 5-bit plaintext letter coding is left to right. Find the polynomial and write it in the form $x^5 + c_4x^4 + c_3x^3 + c_2x^2 + c_1x + 1$.

The Block Cipher Function— A Modern Keystone to the Paradigm

In the last module, we saw that the m-sequence generator could not be made to serve as the fulfillment of a cryptosystem. There was simply too much dependence between the m-sequence bits, and its output could not therefore be considered as a good approximation to Bernoulli behavior. The m-sequence generator was also a linear machine, something that makes its behavior even more easily predictable. In this module, we consider a component that can be used to fulfill the keytext generator paradigm. Cryptosystems can be designed around this component and provide various confidentiality modes for encryption.

Modern *one-key* commercial cryptography is built around a component termed the *block cipher function*. This component is a set of *b* b-input non-linear Boolean functions. The *b* b-input Boolean functions are selected by the secret keying variable *K*. Figure 10-1 sketches what we have diagrammatically.

The block cipher function is crafted so that it possesses the following properties:

- Each output bit $o_j, j = 1, 2, \ldots, b$ is a nonlinear Boolean function of all the b-input bits, $i_1, i_2, \ldots, i_b$. In other words,

$$o_j = f_j(i_1, i_2, \ldots, i_b), j = 1, 2, \ldots, b \tag{10.1}$$

- The block cipher function is reversible. When it is in this state, which is indicated by CIPH_K^{-1}, we can obtain the b-input bits from the b-output bits by putting the b-output bits into the input bits register of the cipher block function. See Figure 10-2.

Figure 10-1
The block cipher function

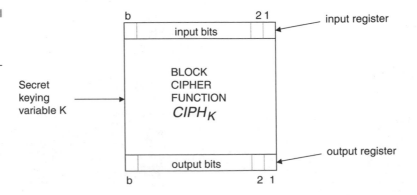

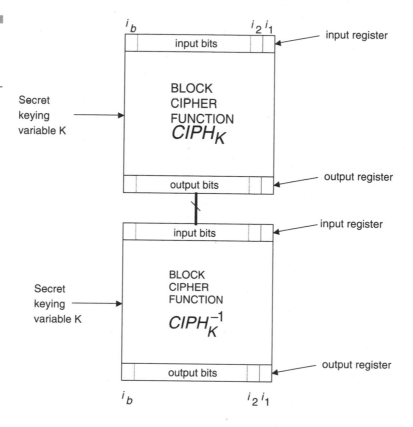

Figure 10-2
The cipher block
function and its
inverse

- For any particular b-bit input, $i_1, i_2, \ldots, i_b$,
    - Changing a single input bit causes each output bit to be inverted with a probability of one-half.
    - Changing a single keying variable bit causes each output bit to be inverted with a probability of one-half.

The first block cipher function endorsed by the Department of Commerce's National Bureau of Standards was the *Data Encryption Algorithm* (DEA) or *Data Encryption Standard* (DES). Published as the *Federal Information Processing Standard Publication* (FIPS PUB) number 46, which was dated 1977, the DES specified a cipher block function with $b = 64$ under the control of a keying variable that is 56 bits long. As time progressed, concerns grew that the DES would not continue to provide sufficient security and two directions were taken from the DES. One of these

Figure 10-3
Triple DES

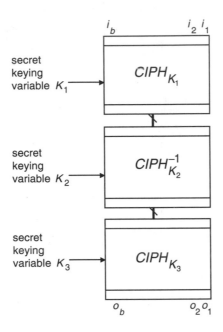

was to concatenate the DES with two copies of itself, as shown in Figure 10-3. This concatenation, which was called *triple DES*, provides much greater security as it can use up to three independent keying variables.

The other direction from the DES was the public solicitation of an *Advanced Encryption Standard* (AES). The candidate accepted was a cipher block algorithm that uses a 128-bit keying variable with b selectable as 128, 192, or 256 bits. The algorithm was published in 2001 as FIPS PUB 197.

Exercise 10

The purpose of this exercise is to examine one of the famous block cipher functions—the DES—and, using what we've learned in Module 7, "Finite State Sequential Machines," see how its inverse can be implemented.

1. The DES, initially promulgated in 1977, is a block cipher function with $b = 64$ bits and a secret keying variable, K, of 56 bits. The algorithm is based on 16 rounds. In a round a 64 block is split into 2 halves, operating on the right half with a complex function and part of the keying variable K. The result is bit-by-bit mod-2 added to the left half and the halves are interchanged—that is, it makes the left half the next right half and the right half the next left half, as indicated in Figure 10-4.

Figure 10-4
The DES encryption process

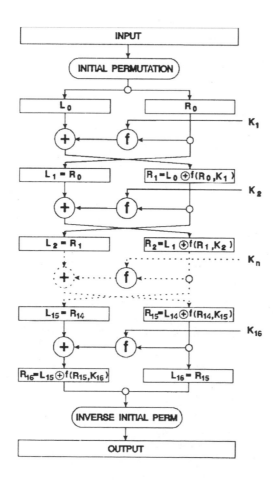

The initial permutations are fixed 64-element permutations. L_i and R_i are, respectively, the left and right 32-bit words that are initialized at $i = 0$ to the two halves of the permuted 64-bit input word. K_i, i = 1, 2, . . . , 16 are each 48 bits that are chosen, according to a fixed pattern, from the 56-bit key. The function f is the heart of the algorithm. It is displayed in Figure 10-5.

The function E is a fixed expander function. Its purpose is to expand the 32 bits of R_{i-1} to 48 bits. It does this by simply repeating 16 of the bits. The function P is a fixed permutation. The modules labeled $S_1, \cdots , S_8$ have come to be called the *S-boxes*. They are all different, but are four six-input nonlinear Boolean functions.

Figure 10-4 illustrates the DES in the encryption mode—in other words, performing what we have labeled CIPH_K. We have said that the block cipher function may be run in the reverse or decryption mode—that is, perform CIPK_K^{-1}. How can this be done? Explain how the structure shown in Figure 10-4 enables this.

2. Figure 10-3 illustrates the structure for triple DES. Why is the second cipher block function operated in the decrypt mode?

Figure 10-5
The complex function
$f = f(R_{i-1}, K_i)$, i = 1, 2,
. . . , 16

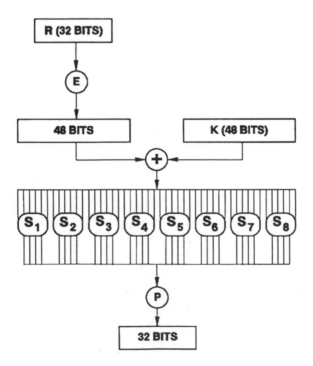

11

Confidentiality Modes: ECB and CTR

Now that we have studied the block cipher function, we can proceed to employ it as the heart of a high-grade commercial cryptographic system, fulfilling the paradigm of Figure 4-1 in Module 4, "Toward a Cryptographic Paradigm." This will be accomplished through a series of confidentiality modes. These are modes or architectures to use with the block cipher function. Each of the modes has distinct characteristics, advantages, and disadvantages. It is important to be aware of and understand the individual mode characteristics in order to select an appropriate one to fulfill your particular cryptographic need.

The first confidentiality mode that we will study is the simplest, most basic method for operating the block cipher function. Termed the *electronic codebook* (ECB) mode, this mode simply uses the block cipher function to encrypt or decrypt a b-bit input word under the control of a secret keying variable *K*. The ECB mode is diagrammed in Figure 11-1.[1]

The ECB mode is one of only two modes that use the block cipher function in the decrypt mode for decryption.

In the ECB mode, the same b-bit input block yields the same b-bit output block for the same secret keying variable *K*. As such, the ECB mode is not a suitable confidentiality mode for the encryption of most message traffic. However, the mode is well suited for encrypting secret keying variables or

Figure 11-1
The ECB mode

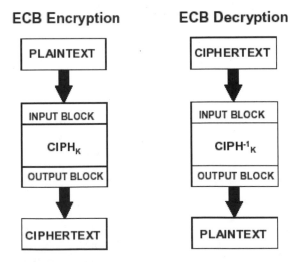

ECB Encryption

PLAINTEXT → INPUT BLOCK / CIPH$_K$ / OUTPUT BLOCK → CIPHERTEXT

ECB Decryption

CIPHERTEXT → INPUT BLOCK / CIPH$^{-1}_K$ / OUTPUT BLOCK → PLAINTEXT

---

[1]From "Recommendation for Block Cipher Modes of Operation: Methods and Techniques," by Morris Dworkin, *NIST Special Publication 800-38A*, 2001 Edition.

other cryptographic variables that must be securely transmitted. It is also suitable for password verification. In this mode, a user generates a password and the password generation function uses the user's password (usually after a publicly known hashing process has been applied to it) as the keying variable for the block cipher function. The password generation function then proceeds to encrypt a b-bit-known string, such as b-zeros, to encrypt the result of the previous encryption, and so on for a plurality of cycles. The result of the last encryption is then stored as the password check. Symbolically, if PT_i is the i-th plaintext, CT_i is the i-th ciphertext, and K_{PW} is the hashed password serving as the secret keying variable for the block cipher function, then

$$CT_i = CIPH_{K_{PW}}(PT_i) \qquad (11.1)$$

and we have

$$\begin{aligned} CT_1 &= CIPH_{K_{PW}}(0 \cdots 0) \\ CT_2 &= CIPH_{K_{PW}}(CT_1) \\ &\vdots \\ CT_N &= CIPH_{K_{PW}}(CT_{N-1}) \end{aligned} \qquad (11.2)$$

The password check, CT_N, is not secret and is stored in a file awaiting comparison to the CT_N generated according to a proffered password.

The ECB mode may also be useful for *key updating*. Key updating is a technique in which a keying variable is operated on using a *one-way function*. We term a function, f, as one way if we are given f and x it is relatively easy to compute $f(x)$, but much more difficult to compute x given f and $f(x)$. Suppose that we update the k-bit block cipher keying variable at time t, $K(t)$, by the following method. We encrypt a b-bit block of all zeros using the ECB mode with the secret keying variable $K(t)$. We select the first k bits of the output block as the new keying variable $K(t + 1)$.[2] The generation of K($t + 1$), $K(t + 2)$, ... is certainly a one-way function.

If we could somehow write down all 2^k values of $K(t)$ and their successor states, $K(t + 1)$, we would find that the correspondence constituted a *mapping* in the sense that we defined the word at the end of Module 7, "Finite

---

[2]We're assuming k ≤ b.

State Sequential Machines;" that is, some values (or states) of $K(t + 1)$ would have no predecessors and some would have multiple predecessors.

M. Kruskal has found a result that sheds a bit of light on this problem.[3] Kruskal found that if a large number of elements (or states) undergo a randomly generated mapping, the mapping will induce *components*. Components differ from cycles, which are induced by a permutation, as a component includes all the states in a cycle plus the lead-in *trees*, which are strings of states that eventually lead to a cyclic state that has more than one predecessor. Machine (c) in Module 7 exhibits components. Note that there are two cycles: one of length 1 and one of length 2. Each cycle has a tree of states leading into it. Kruskal discovered that if the number of states N (in our case, $N = 2^k$) is large, then the average number of components (N_{COMP}) generated by a random mapping is

$$N_{COMP} = \frac{1}{2}\ln(N) + 0.289 \tag{11.3}$$

Key updating can be a very useful endeavor. Suppose, for example, we wanted to carry a recorder or electronic notepad to make confidential notes during a multiday business trip and had no secure place to keep our recording instrument. We could load our device with $K(t)$ just before we start off on the trip and leave a copy of $K(t)$ in a secure receptacle at the office. Then, immediately after we record a thought or message (perhaps as we let up on the record button, for example), a circuit inside our recorder performs a key update. Thus, if we lose our device or it somehow falls into the wrong hands, no compromise of the material occurs as the key cannot be backed up. (We could even drop the device into the mail to be sent home to us rather than carry it around after we're done.) When we're back home, we load the stored initial key, $K(t)$, and decrypt the encrypted messages by updating the key every time we have a message break.

The ECB mode, incidentally, owes its name to the *two-part* code, a cryptographic instrument of great importance in World War I. One part of the two-part code enables encrypted entries to be listed alphabetically. The other part enables entries to be placed in a lexicographical listing of the code elements so that they can be quickly located. Table 11-1 is a listing of the first 15 entries in both parts of an *American Expeditionary Force* (AEF)

---

[3] M. Kruskal, "The Expected Number of Components Under a Random Mapping Function," *American Mathematical Monthly* 61 (1954): 392–7.

Table 11-1

Extract of an AEF two-part code

Encrypt Part Plaintext	Code Group	Code Group	Decrypt Part Plaintext
About to advance	BY	AB	Left
Ammunition exhausted	FB	AF	Enemy machine gun fire serious
Are advancing	PX	AG	Gas is being released
At	SX	AP	Stretcher bearers needed
Attack failed	BM	AV	Recall working party
Attack successful	PF	AW	Casualties heavy
Barrage wanted	XF	AX	Using gas shells
Be ready to attack	ZF	AZ	Relief completed
Being relieved	XA	BD	How is everything
Captured	CB	BF	Right
Casualties heavy	AW	BJ	Situation serious
Casualties light	FZ	BM	Attack failed
Center	PB	BP	Enemy trenches
Enemy	FC	BS	Raiders have left
Enemy barrage commenced	PV	BX	Falling back

code created for use in the front-line trenches. In the ECB mode, of course, input and output words are not listed alphabetically, but you may think of all 2^b input words as matched one-to-one against all 2^b output words and as having the capability to go with equal facility from input word to matched output word or output word to matched input word.

The second confidentiality mode that we look at is a relatively new mode called the *counter* (CTR) mode. It is diagrammed in Figure 11-2.[4] The CTR mode is essentially a coupling of a counter or other suitable finite state sequential machine with the ECB mode. The CTR mode is a direct realization of the paradigm of Figure 4-1 in Module 4. In the CTR mode, the block

---

[4]"Recommendation for Block Cipher Modes of Operation: Methods and Techniques," op. cit.

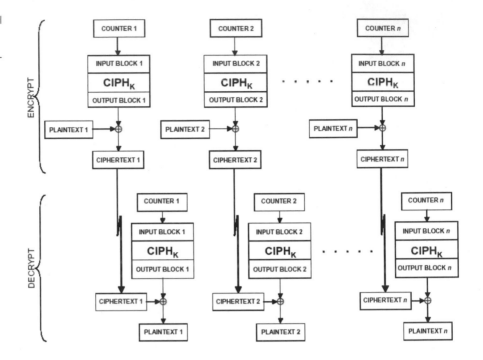

cipher function is operated in the encrypt mode and its input block is filled with a setting that is successively changed. In the case of a counter, the setting is simply incremented. As long as the same setting is not used twice, the keytext generated by the block cipher function will not repeat and cause a depth situation. Figure 11-2 depicts the encryption and decryption process with time moving from left to right. The successive counter inputs to the block cipher function input blocks are labeled as COUNTER 1, COUNTER 2, and so on.

Exercise 11

The purpose of these exercises is to stimulate early thought about crypto-graphic net architecture and develop a probabilistic appreciation for acci-dental depth in one of the confidentiality modes.

1. How could each of the M users form a net safely using the CTR mode to send messages to all the other net members without risking the creation of a depth condition?

2. Suppose that the input to the block cipher function in CTR mode was taken from a publicly known or disclosed random number source. What is the approximate probability that the content of the b-bit input block of the block cipher function will not be repeated in n encryptions, $n \gg 1$?

Confidentiality Mode: Output Feedback (OFB)

The block cipher function can be used in many ways to create a cryptosystem. The mode that we study in this module is well adapted for high-speed operation. It requires an initialization vector *(IV), and its keytext generation is a cyclic process.*

The third confidentiality mode is the *output feedback* (OFB) mode. It uses the block cipher function operated in the encrypt mode, and its starting point is determined by a *b*-bit IV that is loaded into the block cipher function's input register before encryption is commenced. The *b*-bits produced by the first encryption and delivered to the output block are used as keytext and the next contents of the input register. Therefore, the encryption proceeds, using the output block contents as keytext and the IV for the next cycle, as depicted in Figure 12-1.[1] As mentioned before, the encryption and decryption processes are depicted with time moving from left to right.

In theory, if it were operated long enough, the *b*-bit keytext produced would be the same as the IV that the OFB mode started with. This is a con-

Figure 12-1
The OFB mode

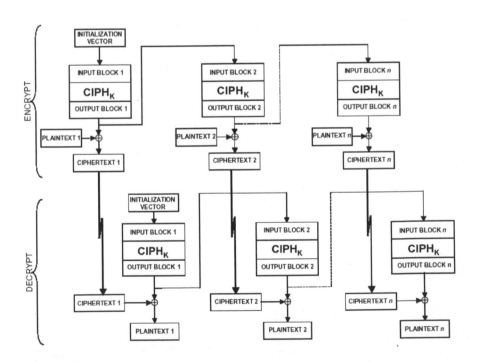

---

[1]From "Recommendation for Block Cipher Modes of Operation: Methods and Techniques," op. cit.

sequence of the OFB mode functioning as a machine that can be characterized as a permutation, such as Machine (d) of Module 7, "Finite State Sequential Machines." However, the cycle length will almost certainly be extremely long, and the probability of a depth condition arising from reusing an entire cycle is extremely small.

Although we cannot know the exact cycle length that we will have from a particular IV in conjunction with the secret keying variable K, we can estimate its length. To do this, we use a valuable discovery by R. Greenwood.[2]

Greenwood explains that if a random one-to-one (permutation) mapping is applied to a set of N elements (N large), then the average number of cycles (N_{CYCLE}) induced will be

$$N_{CYCLE} = \ln(N) + 0.577 \tag{12.1}$$

N_{CYCLE} has a variance, σ^2, of

$$\sigma^2 = \ln(N) - 1.068 \tag{12.2}$$

The average cycle length for the OFB, $\hat{L}$, is found by dividing the total number of states of the input register of the block cipher function, $N = 2^b$, by the expected number of cycles. This results in the following:

$$\hat{L} = \frac{2^b}{b\ln(2) + 0.577} \tag{12.3}$$

Table 12-1 gives $\hat{L}$ for four common values of b.

Table 12-1

$\hat{L}$ versus the input register size (b)

b	$\hat{L}$
64	4.10×10^{17}
128	3.81×10^{36}
192	4.70×10^{55}
256	6.50×10^{74}

[2]R. Greenwood, "The Number of Cycles Associated with the Elements of a Permutation Group," *American Mathematical Monthly* 60 (1953): 407–9.

Exercise 12

The purpose of this exercise is to introduce you to an active attack. Most of the attacks we have considered have been passive wherein the eavesdropper does nothing more than monitor (listen) and analyze encrypted communications. A very different genre of attack is the active attack where the interloper takes an active part in altering, deleting, delaying, or insinuating message traffic. Some active attacks are much more dangerous than passive attacks, but they require a great deal more sophistication on the part of the opposition. Also, it is usually quite difficult to prove a passive attack, whereas an active attack, if compromised, can be spectacular evidence of hostile intent. Quite often, whether or not an active attack succeeds is a function of the cryptographic mode of operation. The following example focuses on this point.

1. Assume that you are relaying an encrypted message from a commander to the headquarters. The structure of the message is shown in Figure 12-2. OFB encryption is used and letters are encoded using the convention in Table 3-1 of Module 3, "Digitization of Plaintext." Assume that you know that the plaintext of the message sent to you for relay is NOT READY. You are an infiltrator and want the message to be decrypted as NOW READY. The ciphertext consists of bits 1 through 45. Can you do this? If so, what bits of the ciphertext do you invert?

Figure 12-2
Message structure

| IV | CIPHERTEXT |

Confidentiality Modes: Cipher Feedback (CFB) and Cipher Block Chaining (CBC)

In this module, we look at two confidentiality modes that can be used with the block cipher function. These modes differ from the previous modes we have studied in a number of ways. The most remarkable of these differences is that the new modes exhibit error extension; that is, errors in transmission not only affect the decryption of the block in which they occur, but they also affect the decryption beyond the block in which they occur.

The first of the remaining two confidentiality modes that we will study is termed the *cipher feedback* (CFB) mode. Like the *output feedback* (OFB) mode, it uses the block cipher operated in the encrypt mode and its starting point is determined by a *b*-bit *initialization vector* (IV) that is loaded into the block cipher function's input register before encryption is commenced.

The CFB mode can be configured in a number of ways depending on the number of bits, s, $l \leq s \leq b$, of the *b*-bit output block that are used for key-text. When s bits of keytext are used, the mode is more specifically termed the *s-bit CFB mode*.

The s bits of ciphertext that result from the combination of the *s*-keytext bits with *s*-plaintext bits become part of the next contents of the input register of the block cipher function. The flow is depicted in Figure 13-1.[1] As mentioned previously, the encryption and decryption processes are depicted with time moving left to right.

By inspecting the dynamic nature in which the input block of the block cipher function is filled, it is clear that if an error occurs in the reception of the ciphertext, the error will propagate through the next $\left\lceil \dfrac{b}{s} \right\rceil$ block cipher function encryptions; thus, the effects of a single error may extend beyond its point of occurrence. This is termed *error extension*. For example, if we are using 1-bit CFB and we have a single error in reception, then the decrypted plaintext bit corresponding to the ciphertext bit received in error will be in error and any of the b bits of decrypted plaintext following that error might also be in error. (The function $\lceil \ \rceil$ is the ceiling function. It specifies the smallest integer that is greater than or equal to its argument.)

Why in the world would we want to use a mode of operation that has such a potentially horrendous error extension? Why not stick with OFB or the *counter* (CTR) mode? The answer to this reasonable question is that there are security and operational trade-offs. Error extension can be a valuable ally in negating some active attacks such as undetected message

---

[1]From "Recommendation for Block Cipher Modes of Operation: Methods and Techniques," op. cit.

Figure 13-1

The CFB mode

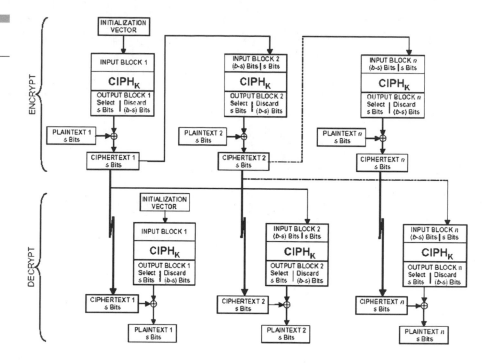

manipulation. Think back to the active attack in Exercise 12. The plaintext could be successfully manipulated because changing the ciphertext did not affect the following keytext. Thus, it is often the case, but not necessarily always the case (as we'll see in an exercise problem), that the CFB mode will deter hostile ciphertext manipulation by rendering the decrypted plaintext unintelligible.

There is also an operational advantage. Consider that in the OFB mode, an error in the IV renders the entire message unintelligible. Also, if the IV is transmitted as the first part of the message, it is impossible for a receiver to start decrypting the message if it fails to capture the IV for any reason, including preamble jamming. However, with CFB mode, because the input register is filled with transmitted ciphertext, the input register will eventually, barring reception errors, contain the correct contents to generate the keytext necessary to correctly decipher plaintext. For example, in 1-bit CFB mode, we need wait on only b consecutive bits of error-free ciphertext before we begin to correctly decrypt the subsequent ciphertext.

Error extension (by nature) and the CFB mode (by implication) impose an operational disadvantage. Additional power or the signal-to-noise ratio

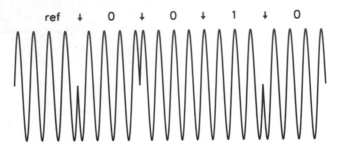

Figure 13-2
The decoding of a
received DPSK signal

must be maintained in order to produce decrypted plaintext at an error rate
that is comparable to that produced using the OFB mode.

We contrast the OFB mode with the 1-bit CFB mode in the case of *differential phase-shift keying* (DPSK). DPSK does not require the receiver to
have a coherent reference signal, but rather that it measure only relative
phase differences.

Figure 13-2 illustrates the decoding of a received DPSK signal. We have
used arrows to indicate the points at which the signal phase of the carrier
may be reversed. When the carrier phase is reversed, the decoded symbol is
a zero; when the phase is not reversed, the decoded symbol is a one.

The bit error rate, p_e, for a DPSK transmission received in additive white
Gaussian noise is

$$P_e = \frac{1}{2}e^{-E_b/N_o} \qquad (13.1)$$

where E_b/N_0 is the signal-to-noise ratio, which is usually expressed in *decibels* (dBs). The effect of the error extension can be factored into a margin
for the signal-to-noise ratio that is necessary to achieve an error rate
corresponding to a confidentiality mode that has no error extension, such
as OFB.

Figure 13-3 plots the effects of error extension on p_e for a DPSK transmission for two different values of the block length (b). Note that selecting
the CFB mode can cost us close to a couple of dBs over the OFB mode.

Why might we be willing to pay such a price? One reason has to do with
the ease of synchronization. If you are operating a mobile network with
short-lived fades, for example, you may occasionally miss the beginning of
a transmission. With OFB, you're out of luck as the OFB mode requires an

Figure 13-3

The effect of error extension on a DPSK signal[2]

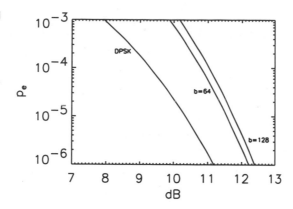

error-free IV. However, with 1-bit CFB, after b consecutive bits have been received without error, the receiver will begin to properly decipher the enciphered message.

The final confidentiality mode is the *cipher block chaining* (CBC) mode. Like the CFB mode, it introduces error extension, but the characteristics of the extension are a bit different. In a way, it is somewhat similar to the *electronic codebook* (ECB) mode as it uses the block cipher function in the decrypt mode for decryption. The flow of the mode is depicted in Figure 13-4.[3]

---

[2]Just a short note for those of you interested in *radio frequency* (RF) communications. There is an intuitive tendency to believe that DPSK will produce errors that are primarily doublets. In additive white Gaussian noise, this is not necessarily true. This has been noted and can be found in J. Salz and B. Saltzberg's paper, "Double Error Rates in Differentially Coherent Phase Systems," *IEEE Transactions on Communications Systems* (1964): 202–5: "In digital communications systems using binary phase modulation and differentially coherent detectors, it seems reasonable to suspect that errors tend to occur in pairs. The intuitive arguments are plausible, since these receivers obtain their reference from past bits. . . . It is shown that when the cause of errors is additive Gaussian noise the ratio of double errors to single errors is less than 25 percent when the SNR [Signal-to-Noise Ratio] is larger than 5 dB. As a matter of fact, this ratio approaches zero as the SNR increases without bound. . . . [For] isolated impulses as the disturbance instead of Gaussian noise . . . most errors [occur] in isolated pairs."

[3]From "Recommendations for Block Cipher Modes of Operation: Methods and Techniques," op, cit.

Figure 13-4
The CBC mode

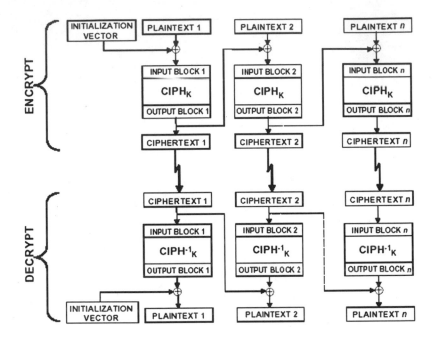

Recall that a deliberate change in the ciphertext of a message protected under the OFB confidentiality mode results in a change in the corresponding portions of decrypted plaintext. Note that there is a vulnerability analogy in the CBC mode in that a deliberate change introduced into the IV causes a change in the corresponding portion of the first ciphertext block.

Exercise 13

In this exercise, you'll look at questions concerning error extension. Examine them carefully (especially the second question), do your best, and then check the answers. Think about what the result would be under other sets of parameters.

1. Two block cipher functions are operated in the 1-bit CFB mode, as shown in Figure 13-5. Note that one function operates on b-bit words and the other operates on n-bit words.

   An error is made in the transmission of the i-th ciphertext bit $ct(i)$. What is the largest value, j, for which the decrypted plaintext, $pt(i + j)$, might be in error?

2. The *Data Encryption Standard* (DES) is operated in 64-bit CFB mode. The following 25-character message is sent:

   `HAVE_RECEIVED_ONE_MILLION`

   The _ character is the space character. Assume that the characters are encoded using the plaintext coding table in Figure 3-1 of Module 3, "Digitization of Plaintext." Assume also that the plaintext of the message you are sending is known to your adversary who has access to your ciphertext and can alter it in an active attack. Your adversary wants to alter the ciphertext of your message so that it decrypts to

   `HAVE_RECEIVED_TWO_MILLION`

   Can your adversary do this?

Figure 13-5
Two block cipher
functions operating
in tandem

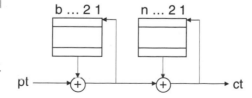

II

Number Theory We Will Need

This second part comprises eight modules. These modules identify and teach a few principles from one of the most vast and fascinating disciplines of mathematics, number theory.

Number theory can consume the academic life of a mathematician. This particular branch of mathematics is rich and deep, and it is by no means an easy task to extract just a few pages and call it representative or complete unto itself. But we have to start somewhere, and I believe that these few topics are essential for an introduction to public key cryptography. Please take your time and become comfortable with the concepts. You won't find them too foreign. I expect that they will seem strangely familiar as many of the results of number theory touch us at so many points of our professional lives.

Prime Numbers I

"The integers were created by God. Everything else is man's work," said Leopold Kronecker, the famous mathmatician. In this module, we study a subset of the integers called natural numbers *and an important subset of those numbers called* prime numbers. *Prime numbers are central to much of public key cryptography, and it is essential that we develop an almost visceral feel for elementary number theory. Primes are a natural place to start.*

Integers are whole numbers that run from minus infinity to plus infinity, that is . . . , $-3, -2, -1, 0, 1, 2, 3, \ldots$. Natural numbers are the positive integers $1, 2, 3, \ldots$. A prime is a natural number that has exactly two distinct divisors, such as $2, 3, 5, 7, 11, 13, \ldots$. There are 25 primes that are less than 100 and 168 primes that are less than 1,000.

A *composite number* is a natural number that has more than two distinct divisors, such as $4, 6, 8, 9, 10, 12, \ldots$.

This leaves the number one, or *unity*. One is neither a prime nor composite number. It is its own class.

The *canonical decomposition* of a number (N) is

$$N = \prod_i p_i^{\alpha_i} \tag{14.1}$$

where $\{p_1, p_2, p_3, \ldots\}$ are all the primes and $\{\alpha_1, \alpha_2, \alpha_3, \ldots\}$ are the exponents of the primes. Thus,

$$\begin{aligned} 6 &= 2^1 \times 3^1 \\ 484 &= 2^2 \times 11^2 \end{aligned} \tag{14.2}$$

For unity, the $\{\alpha_i\}$ are all zero, which is why number theorists refer to unity as *the empty product of primes*.

The canonical decomposition in Equation 14.1 is unique for every N. This fact has been given an imposing name: the *Fundamental Theorem of Arithmetic*.

How many primes are there? This is an old question that has an old answer. Tradition holds that Euclid proved that there is an infinity of prime numbers. The argument is quite clever. Assume we know all of the prime numbers—$p_1, p_2, \ldots, p_n$. We form the product of all the primes and add unity; in other words, we form the number π_p where

$$\pi_p = 1 + \prod_{i=1}^{n} p_i \tag{14.3}$$

It is clear that none of the known primes divide π_p so we must conclude that either

- π_p is a previously unknown prime.
- π_p is the product of previously unknown primes.

Either way, we see that we cannot bound the number of primes as finite.

The density of primes becomes increasingly thin as N increases. We already noted that the number of primes less than 1,000 was less than 10 times the number of primes less than 100. Furthermore, there are stretches of unlimited-length successive numbers that are entirely prime free. Again, the proof is very simple. Consider the stretch of successive numbers:

$$n! + 2, n! + 3, \ldots, n! + n \tag{14.4}$$

The interval in Equation 14.4 is $n - 1$ numbers long and must be prime free. Why?

The celebrated *Prime Number Theorem* of vintage 1896 disclosed that as $N \to \infty$, the number of primes that do not exceed N is approximated by $N/\ln(N)$. So, we will never run out of primes and, even though their density becomes increasingly thin as N increases, many m-digit primes are still available for large m.

One of the most important tools of elementary number theory is the *Euclidean Algorithm*. This algorithm finds the *greatest common denominator* (GCD) of two numbers, n_1 and n_2, and is often written as (n_1, n_2). The GCD of n_1 and n_2 is the largest number, d, that divides both n_1 and n_2. The GCD is calculated for n_1 and n_2 by the algorithm shown in Figure 14-1. (We assume that $n_1 > n_2$.)

Let's try the Euclidean Algorithm with $n_1 = 341$ and $n_1 = 231$. Table 14-1 shows the calculations.

If n_1 and n_2 are independently chosen from a uniform distribution over the interval $[1, N]$ for N large, then this form of the Euclidean Algorithm will require about $0.843\ln(N)$ replications on average and a maximum of about $2.078\ln(N)$ replications.

If you need to build circuits out of elementary cells to perform the Euclidean Algorithm, you may want to use the variant shown in Figure 14-2, which only requires division by two, which is, of course, easily done by a simple 1-bit position shift. However, we must first do a bit of preconditioning. We will assume in Figure 14-2 that both n_1 and n_2 are odd. If they are not both odd, two possible cases exist and they should be handled as follows:

- If both n_1 and n_2 are even, successively divide both of them by two until at least one of them is odd. Count the number of times that you

Figure 14-1
The Euclidean
Algorithm

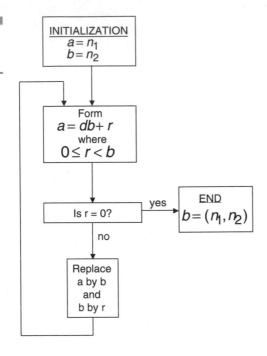

Table 14-1

The flow of the
Euclidean
Algorithm of
Figure 14-1

a	b	d	r
341	231	—	—
341	231	1	110
231	110	—	—
231	110	2	11
110	11	—	—
110	11	10	0

divide by two because this number must be incorporated as part of the GCD of n_1 and n_2.

■ If exactly one of n_1 and n_2 is even, successively divide the even number by two until it is odd.

Let's also try this version of the Euclidean Algorithm with $n_1 = 341$ and $n_2 = 231$. Table 14-2 shows the calculations.

Figure 14-2
The Euclidean Algorithm with division limited to division by two

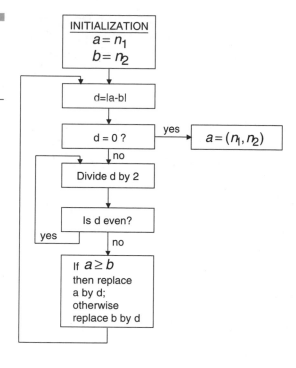

Table 14-2

The flow of the Euclidean Algorithm of Figure 14-2

a	b	d
341	231	—
341	231	110
341	231	55
55	231	176
55	231	88
55	231	44
55	231	22
55	231	11
55	11	—
55	11	44
55	11	22
55	11	11
11	11	0

If n_1 and n_2 are independently chosen from a uniform distribution over the interval $[1, N]$ for N large, then this form of the Euclidean Algorithm requires about $1.019\ln(N)$ replications on average and a maximum of about $1.443\ln(N)$ replications.[1]

If $(n_1, n_2) = 1$, we say that n_1 and n_2 are *relatively prime*. Note that 14 and 15 are relatively prime yet neither 14 nor 15 is a prime number.

We must discuss one more definition and relation. The *least common multiple* (LCM) of n_1 and n_2 is written (n_1, n_2) and is defined as the smallest number that is divisible by both n_1 and n_2. The GCD and LCM are related, as shown in Equation 14-5.

$$\{n_1, n_2\} = \frac{n_1 \times n_2}{(n_1, n_2)} \tag{14.5}$$

[1] R. Brent, "Analysis of the Binary Euclidean Algorithm," in (Symposium on) *Algorithms and Complexity*, J. Traub, ed., (Academic Press, 1976), 321–55.

Exercise 14

This exercise checks your proficiency with the nuts and bolts of the Euclidean Algorithm and also introduces you to an interesting, special class of numbers.

1. Find the GCD of 1,496 and 1,989.

2. The set of natural numbers, $N = \{1, 2, 3, 4, \ldots\}$, consists of three types of numbers:

   - Unity
   - Prime numbers
   - Composite numbers

   The Fundamental Theorem of Arithmetic states that any natural number is expressible in the canonical decomposition $N = p_1^{\alpha_1} p_2^{\alpha_2} p_3^{\alpha_3} \ldots$ where $p_1, p_2, p_3, \ldots$ are the primes and $\alpha_1, \alpha_2, \alpha_3, \ldots$ are the exponents of the primes. This expression is unique. Express 100 in its canonical decomposition.

   Now consider that we are dealing only with those natural numbers that leave a remainder of unity when divided by three. Let this set be denoted by S and $S = \{1, 4, 7, 10, \ldots\}$. Note that the product of any two numbers in S is also in S. We define a prime number in S as a number that has exactly two divisors that are in S; for example, 7 and 22 are both primes in S. You might wonder if the canonical decomposition of a member of S is unique. The integer 100 is a member of S. Find all the canonical decompositions of 100 using only the primes found in S.

Congruences

Congruences, which involve the study of congruential relations, are an essential element of number theory and figure prominently in our study of public key cryptography. It is important that you know and are comfortable with the basics of this topic. You will find that the concept of congruences follows a natural way of thinking, and their operations and terminology will become second nature.

If the number m divides $a-b$, written in shorthand as $m|(a-b)$, then we say that a is *congruent* to b *modulo* (*mod*) the number m. This is equivalent to saying that $a-b$ is a multiple of m. If a is congruent to b *mod* m we write $a = b \ mod(m)$

Congruences are all about remainders or residues resulting from dividing by a modulus m. If we divide a positive number by the modulus m, the remainder r is in the following set of numbers:

$$\{0, 1, 2, \ldots, m-2, m-1\} \tag{15.1}$$

If we divide a negative number by the modulus, the remainder is either in the set in 15.1 or can be brought into the set, or reduced, by adding the modulus to the remainder. For example, if $m = 10$, the set in 15.1 becomes

$$\{0, 1, 2, 3, 4, 5, 6, 7, 8, 9\} \tag{15.2}$$

Consider that we have the number 37. If we divide by the modulus, 10, we have a remainder of 7. Note that 7 is in the set in 15.2. Suppose we started with -37. Dividing by 10 yields a remainder of -7. Now -7 is not within the set in 15.2, but if we add the modulus (10) to -7, we get 3, which is in the set in 15.2. We say that 37 is congruent to 7 mod 10 and -37 is congruent to 3 mod 10.

The set in 15.1 has a very special place in elementary number theory and a special name. It is denoted by R and is called the *principal complete residue set mod m*. Every number is congruent mod m to a unique member of R.

The principal complete residue set has a number of important properties. First, if the number b is added to every member of the principal complete residue set, the m numbers produced by the addition and reduced so that they appear in 15.1, also form a principal complete residue set mod m. (The set may be permuted, but all the elements are there.) For example, let's pick $b = 4$ and add it to the members of the principal complete residue set mod 10. We get 15.3.

$$\{4, 5, 6, 7, 8, 9, 0, 1, 2, 3\} \tag{15.3}$$

A second important property concerns multiplication. If we multiply each member of a principal complete residue set mod m by a number a, where a and m are relatively prime, that is, $(a, m) = 1$, then we will also obtain a principal complete residue set. For example, let's pick $a = 7$ and $m = 10$. The result is shown in 15.4.

$$\{0, 7, 4, 1, 8, 5, 2, 9, 6, 3\} \tag{15.4}$$

Note that 15.4 is the principal complete residue set mod 10; the elements are just permuted. What would have happened if $(a, m) \neq 1$? Say we had chosen $a = 6$. By multiplying each member of the principal complete residue set for $m = 10$, we get 15.5.

$$\{0, 6, 2, 8, 4, 0, 6, 2, 8, 4\} \tag{15.5}$$

The result is not a principal complete residue set mod 10.

From the offset, it should be clear that if x spans, that is, takes on the value of each member of, a principal complete residue set mod m, then $ax + b$ also spans a principal complete residue set mod m so long as $(a, m) = 1$.

There is another residue set that is extremely important. This residue set is called the *reduced residue set mod m*. It is denoted by Q and consists of those elements of R that are relatively prime to the modulus m. For the residue set mod 10,

$$Q = \{1, 3, 7, 9\} \tag{15.6}$$

The number of elements in Q, the *cardinality* of Q, will turn out to be a very special number theoretic function called the *phi function*. It can be defined as the summation of an indicator function, as written in Equation 15.7.

$$\begin{aligned} \phi(m) &= \Sigma_1 \\ (a, m) &= 1 \\ 0 < a &\leq m \end{aligned} \tag{15.7}$$

It is important that we know some properties of congruences with respect to the basic operations of multiplication, exponentiation, and division. Many of these properties are essential for performing operations in public key cryptography.

The first set, those properties dealing with multiplication and exponentiation, is as follows:

- If $a \equiv b \bmod(m)$ and we have $c > 0$, then we have $ac \equiv bc \bmod(m)$ and $a^c \equiv b^c \bmod(m)$.

- If we have the set of k equations over k moduli,

  - $a \equiv b \bmod(m_1)$
  - $a \equiv b \bmod(m_2)$
  - $\vdots$
  - $a \equiv b \bmod(m_k)$

then $a \equiv b \bmod\{m_1, m_2, \ldots, m_k\}$ where $\{\cdot, \cdot, \ldots, \cdot\}$ is the *least common multiple* (LCM) of its arguments.

The second set of properties of congruence relations comprises properties dealing with division. Here we have the following:

- If $ac \equiv bc \bmod(m)$ and $(c, m) = 1$, then we may divide both sides of the congruence relation by c and know that $a \equiv b \bmod(m)$.

- If $ac \equiv bc \bmod(m)$ and $(c, m) = d$, then we can still get rid of c, but we have to adjust the modulus of the resulting congruence relation as

  $a \equiv b \bmod\left(\dfrac{m}{d}\right).$

- Finally, if k is a positive divisor of the modulus m and if $a \equiv b \bmod(m)$, then we may write $a \equiv b \bmod(k)$.

As an example of this last relation, we know that $3 \equiv 13 \bmod(10)$. Because $2 \mid 10$ and $5 \mid 10$, we may write $3 \equiv 13 \bmod(2)$ and $3 \equiv 13 \bmod(5)$.

There is one further point we must make about congruences before moving on and that concerns *congruential equations*. When we write

$$f(x) \equiv 0 \bmod(m) \tag{15.8}$$

we are seeking to find all solutions to Equation 15.8 as x spans 15.1, the principal complete residue set mod m. Equation 15.8 may have no solutions, a unique solution, or multiple solutions. For example, let's say we are interested in finding an x in 15.2 such that

$$x^2 \equiv b \bmod(10) \tag{15.9}$$

We set

$$f(x) = x^2 - b \qquad (15.10)$$

and let x span 15.2. We get the set of numbers shown in 15.11.

$$\{-b, 1-b, 4-b, 9-b, 6-b, 5-b, 6-b, 9-b, 4-b, 1-b\} \qquad (15.11)$$

From 15.11, we can easily identify and count the solutions to Equation 15.10 for any b. The results are summarized in Table 15-1.

Table 15-1

Solution sets for
$x^2 - b \equiv 0(10)$

b	Number of Solutions	Solutions
0	1	$x = 0$
1	2	$x = 1$ and $x = 9$
2	0	—
3	0	—
4	2	$x = 2$ and $x = 8$
5	1	$x = 5$
6	2	$x = 4$ and $x = 6$
7	0	—
8	0	—
9	2	$x = 3$ and $x = 7$

Exercise 15

The purpose of this exercise is to help you begin to think better congruentially. It takes a little getting used to oddly enough because it is a lot simpler than what you've been used to doing in arithmetic. When you work modulo m, there are only m distinct numbers. This opens up a basket of shortcuts you wouldn't dream of taking otherwise.

1. Does $3x \equiv 7 \bmod(12)$ have a solution? Does $6x \equiv 3 \bmod(15)$ have a solution?

In problems 2 through 4, express your answer as a member of the principal complete residue set of the modulus.

2. What is $47^{1395} \bmod(48)$?

3. What is $4^{3207} \bmod(1024)$?

4. What is $2^{57} \bmod(123)$?

Euler-Fermat
Theorem

We have now learned enough to develop an important theorem of multiplicative number theory. This theorem gives us insight and the ability to navigate more successfully in public key cryptography.

We have just learned about residue sets, both principal complete and reduced. In this development, we are concerned with the reduced residue set and some of its remarkable properties. As mentioned before, we let Q stand for the reduced residue set mod m. Q has the following distinct elements:

$$Q = \{e_1, e_2, \ldots, e_{\phi(m)}\} \tag{16.1}$$

We multiply each element of Q by any number (a) that is relatively prime to the modulus m—that is, $(a, m) = 1$. This produces the set Q^* where

$$Q^* = \{ae_1, ae_2, \ldots, ae_{\phi(m)}\} \tag{16.2}$$

We assume that the elements in Q^* have been reduced modulo m so that they remain in the range $[0, m - 1]$.

Could Q^* also be a reduced residue set mod m? Indeed, it can and this can be explained by a simple argument by contradiction. First, each element in Q^* is relatively prime to m. This is clearly the case because any element of Q, e_i, is relatively prime to m and $(a, m) = 1$; therefore, $(ae_i, m) = 1$. For the second step, we assume that two distinct elements in Q^* are identical; we will call them ae_i and ae_j. If this is the case, then

$$ae_i \equiv ae_j \, \text{mod}(m) \tag{16.3}$$

We can simply divide both sides of the congruence by a and the contradiction appears.

So, we have found that Q^* is also a reduced residue set mod m; therefore, the product of the elements in Q should be the same as the product of the elements in Q^*. In other words,

$$e_1 e_2 \cdots e_{\phi(m)} \equiv (ae_1)(ae_2) \cdots (ae_{\phi(m)}) \, \text{mod}(m) \tag{16.4}$$

We can then write

$$e_1 e_2 \cdots e_{\phi(m)} \equiv a^{\phi(m)} e_1 e_2 \cdots e_{\phi(m)} \, \text{mod}(m) \tag{16.5}$$

Because all of the $\{e_i\}$ are relatively prime to m, we can divide both sides of the congruence in Equation 16.5 by their product. We have

$$a^{\phi(m)} \equiv 1 \bmod(m) \qquad (16.6)$$

The formula shown in Equation 16.6 is called the *Euler-Fermat Theorem*. The Euler-Fermat Theorem immediately hints at a rich algebraic structure because if we split up the left side so that we have

$$a \times a^{\phi(m)-1} \equiv 1 \bmod(m) \qquad (16.7)$$

or

$$a^{-1} \equiv a^{\phi(m)-1} \bmod(m) \qquad (16.8)$$

we realize that we have an inverse mod m for the element a, so long as a is relatively prime to m.

This structure enables us to discover that the elements of Q constitute an Abelian group under multiplication mod m.

- First, for any two elements of Q, e_i, $e_j \epsilon Q$, $e_i e_j \epsilon Q$ disclosing closure.
- Second, for any three elements of Q, e_i, e_j, $e_k \epsilon Q$, we have $e_i(e_j e_k) = (e_i e_j)e_k$. This is the requirement for associativity.
- Third, Q possesses an operational identity element, $e_1 \epsilon Q$, such that for any element $e_i \epsilon G$, $e_i e_1 = e_i$. (Note that $e_1 \equiv 1 \bmod(m)$.)
- Fourth, every element of G, $e_i \epsilon G$, has an inverse called e_i^{-1} such that $e_i e_i^{-1} = e_1$. The inverse may be found via Equation 16.8.
- Finally, because $e_i e_j \equiv e_j e_i \bmod(m)$, the group is commutative or Abelian.

We are now in a position to prove a famous theorem. The theorem is known as *Wilson's theorem*, and it is one of the few *if-and-only-if* theorems for prime numbers. The theorem states that a number p, $p > 1$, is a prime if and only if

$$p \,|\, 1 + (p - 1)! \qquad (16.9)$$

The first thing we must do is dispose of a specific case. That case is $p = 2$. Clearly, p divides $1 + (2 - 1)!$ Therefore, Equation 16.9 is true for $p = 2$.

The next thing we do is ask for the solution to the following equation:

$$x^2 \equiv 1 \bmod(p) \tag{16.10}$$

We may write Equation 16.10 in a factored form:

$$(x - 1)(x + 1) \equiv 0 \bmod(p) \tag{16.11}$$

An odd prime p satisfying Equation 16.11 must divide $x - 1$ or $x + 1$, but not both. Therefore, from the two equations

$$x - 1 \equiv 0 \bmod(p) \tag{16.12a}$$

and

$$x + 1 \equiv 0 \bmod(p) \tag{16.12b}$$

we see that only 1 and $p - 1$ are their own inverse mod p. This is a useful observation because it means that every element in the reduced residue set, Q,

$$Q = \{1, 2, \ldots, p - 2, p - 1\} \tag{16.13}$$

has its inverse $\bmod(p)$ also in Q except for 1 and $p - 1$. Therefore, if we form the product of all the elements in Q, that is, form $(p - 1)!$, the product $\bmod(p)$ will be only $p - 1$. Additionally, we may write $p - 1$ as $-1 \bmod(p)$. Therefore, if p is an odd prime, we have established the necessity that

$$(p - 1)! \equiv -1 \bmod(p) \tag{16.14a}$$

or

$$p \mid 1 + (p - 1)! \tag{16.14b}$$

Now let's move to sufficiency. Suppose p is not prime and $p > 1$. Then $p = p_1 p_2$ where neither p_1 nor p_2 is 1. Then if $p \mid 1 + (p - 1)!$, so must p_1. But, because $1 < p_1 < p$, $p_1 \mid (p - 1)!$ and therefore cannot divide $1 + (p - 1)!$, proving the theorem.

Wilson's theorem is more an item of curiosity than something that can help us as it certainly wouldn't be a good test for primality as the factorial grows awfully big awfully fast.

Exercise 16

The purpose of this exercise is to use the Euler-Fermat theorem to guide us in computing an inverse. In Module 19, "The Extended Euclidean Algorithm," we calculate the same element inverse by a different method.

1. Find the inverse of 3 mod(101).

The Euler Phi (φ) Function

In this short module, we examine the most important number theoretic function in cryptography: the Euler phi (ϕ) function. We have already defined the Euler ϕ function. We did it in Module 15, "Congruences," where we defined

$$\phi(m) = \sum_{\substack{(a, m) = 1 \\ 0 < a \leq m}} 1 \qquad (17.1)$$

The Euler ϕ function turned out to be extremely important in Module 16, "Euler-Fermat Theorem," where it figures prominently in the inverse of an element $a \bmod m$ where $(a, m) = 1$ as

$$a^{-1} \equiv a^{\phi(m)-1} \bmod(m) \qquad (17.2)$$

This will be a critical concept when we move into the RSA public key cryptographic system and we will have to know how to calculate ϕ.

First, we must present a few simple observations. If p is a prime, then what is $\phi(p)$? To answer this, let's look at the reduced residue set mod p and count its members. The answer is $\phi(p)$, of course, and the reduced residue set for p is

$$(1, 2, 3, \ldots, p - 2, p - 1) \qquad (17.3)$$

It also has $p - 1$ members. Therefore, for a prime p,

$$\phi(p) = p - 1 \qquad (17.4)$$

Now suppose we ask for $\phi(p^n)$ where p is a prime. How many members are in the reduced residue set mod p^n? This can be answered easily if you remember that the only numbers between 1 and p^n that are not relatively prime to the prime p are multiples of p. Therefore, the reduced residue set mod p^n is

$$\{1, 2, \ldots, p - 1, p + 1, \ldots 2p - 1, 2p + 1, \ldots p^k + 1, \ldots, p^n - 1\} \quad (17.5)$$

which has $p^{n-1}(p - 1)$ members; therefore, for a prime p,

$$\phi(p^n) = p^{n-1}(p-1) \tag{17.6}$$

Although we will not prove it, it is not difficult to show that $\phi(\cdot)$ is a *weakly multiplicative function*. This means that

$$\phi(ab) = \phi(a)\phi(b) \tag{17.7}$$

if a and b are relatively prime—that is, if $(a, b) = 1$.

Given that ϕ is a weakly multiplicative function and using Equation 17.6, we can calculate ϕ of any number (m), so long as we can factor m into its canonical decomposition. Thus, if

$$m = p_1^{\alpha_1} p_2^{\alpha_2} p_3^{\alpha_3} \cdots \tag{17.8}$$

for nonzero $\{\alpha_i\}$, then

$$\phi(m) = p_1^{\alpha_1 - 1}(p_1 - 1) \times p_2^{\alpha_2 - 1}(p_2 - 1) \times p_3^{\alpha_3 - 1}(p_3 - 1) \cdots \tag{17.9}$$

The function $\phi(\cdot)$ crops up in many places outside of public key cryptography. As mentioned in Module 8, "m-sequences," there is more than one primitive polynomial for each n > 2. For example, $n = 3$ has two: $x^3 + x + 1$ and $x^3 + x^2 + 1$. The number of primitive polynomials for degree n is

$$\frac{\phi(2^n - 1)}{n} \tag{17.10}$$

Thus, for $n = 3$,

$$\frac{\phi(2^3 - 1)}{3} = \frac{\phi(7)}{3} \tag{17.11}$$

$$= \frac{6}{3}$$

$$= 2$$

Exercise 17

This exercise reveals some further properties implied and possessed by the ϕ function. We also establish a follow-on theorem to the Euler-Fermat Theorem.

1. For the equation

$$\phi(x) = y \qquad (17.12)$$

$y = 1$ has two solutions: $x = 1$ and $x = 2$. For which of the following cases are there solutions?

a. $y = 2$
b. $y = 8$
c. $y = 29$

2. Calculate $\phi(45)$.

3. Calculate the sum of the ϕ function for each divisor of 45—in other words, calculate

$$y = \sum_{d|45} \phi(d) \qquad (17.13)$$

Make an astute observation and guess.

4. p is a prime. $(x + y)^p \equiv (a^b + c^d) \bmod(p)$. Find $a, b, c,$ and d.

The Binary Exponentiation Algorithm

Exponentiation is required to implement the public key cryptographic systems that we will be using. The binary exponentiation algorithm *will be quite useful.*

In this module, we learn how to use exponentiation—that is, efficiently calculate α^n given α and n. The procedure used to do this can be performed with the binary exponentiation algorithm. The algorithm is defined in Figure 18-1. $\lfloor x \rfloor$ is the *floor function*. It is equal to the largest integer in x—that is, the largest integer less than or equal to its argument. So, for example, $\lfloor 5 \rfloor = 5, \lfloor 6.7 \rfloor = 6$, and $\lfloor -4.3 \rfloor = -5$.

The binary exponentiation algorithm can be used for modular or non-modular exponentiation, of numbers, or polynomials as we will see later. It requires

$$\lfloor \log_2 n \rfloor + \sigma(n) \tag{18.1}$$

multiplications, where $\sigma(n)$ is the number of ones in the binary representation of n.

Let's follow an example all the way through. Using $n = 15$, we have the flow recorded in Table 18-1.

We note that calculating α^{15} using the binary exponentiation algorithm requires seven multiplications, as predicted by Equation 18-1. As we have

Figure 18-1

The binary exponentiation algorithm

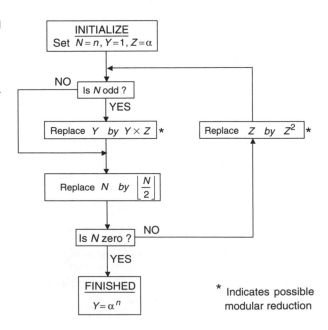

Table 18-1

The binary exponentiation algorithm calculating α^{15}

N	Y	Z	**Running Total of Multiplications**
15	1	α	0
15	α	α	1
7	α	α	1
7	α	α^2	2
7	α^3	α^2	3
3	α^3	α^2	3
3	α^3	α^4	4
3	α^7	α^4	5
1	α^7	α^4	5
1	α^7	α^8	6
1	α^{15}	α^8	7
0	α^{15}	α^8	7

said, the algorithm is efficient, but it is not necessarily the fastest algorithm for a given n. A classic counterexample is given by the following process:

- Set β equal to α.
- Replace β with β^2 (one multiplication).
- Replace β with $\beta \times \alpha$ (one multiplication).
- Set γ equal to β (this saves α^3).
- Replace β with β^2 (one multiplication).
- Replace β with β^2 (one multiplication).
- Replace β with $\beta \times \gamma$ (one multiplication); β now contains α^{15}.

Exercise 18

This exercise asks you to perform the binary exponentiation algorithm using modular arithmetic in preparation for our introduction to public key cryptography. We suggest that your computations be done in the style of Table 18-1, using asterisks to mark modular reductions.

1. Calculate $17^{87} \bmod(101)$.

The Extended Euclidean Algorithm

In Module 14, "Prime Numbers I," we looked at the Euclidean algorithm, which found (a, b), *the greatest common divisor of* a *and* b. *In this short module, we discuss how to use an extension of the Euclidean algorithm—logically termed the extended Euclidean algorithm—to compute some useful associates of the greatest common divisor.*

There exist integers s_n and t_n such that

$$(a, b) = s_n a + t_n b \tag{19.1}$$

The extended Euclidean algorithm provides a way to find s_n and t_n and the inverse of an integer mod(m) if it exists. We'll present the algorithm as it is described by K. Rosen.[1]

We define

$$r_0 = a$$
$$r_1 = b \tag{19.2}$$

and some auxiliary variables

$$d_{j+1} = \left\lfloor \frac{r_j}{r_{j+1}} \right\rfloor \tag{19.3}$$

where

$$r_{j+2} = r_j - d_{j+1} r_{j+1} \tag{19.4}$$

and s_j and t_j where

$$s_j = s_{j-2} - d_{j-1} s_{j-1} \tag{19.5}$$

$$t_j = t_{j-2} - d_{j-1} t_{j-1} \tag{19.6}$$

---

[1]K. Rosen, *Elementary Number Theory*, 4th edition, (Addison Wesley Longman, Inc., 2000).

s_j and t_j are initialized according to

$$s_0 = 1$$
$$s_1 = 0$$
$$t_0 = 0$$
$$t_1 = 1 \qquad (19.7)$$

The flow is just like that shown in Table 14-1 of Module 14 except that we have condensed it a bit, have not repeated the lines, have added an index column for the step indexed by j and two auxiliary variables s_j and t_j, and have extended the algorithm by one step. This all becomes clear when we redo the example of Table 14-1 of Module 14 wherein $a = 341$ and $b = 231$. Table 19-1 shows the flow.

Notice that we have gone one step beyond determining the greatest common divisor, which is r_{j+1} when r_{j+2} becomes zero. The values of s_j and t_j at the next step (here $j = 3$) are the values required in Equation 19.1 as

$$(341,231) = (-2) \times 341 + (3) \times 231 = 11 \qquad (19.8)$$

There is an additional bonus to the extended Euclidean algorithm. It gives you the ability to find an inverse of an element mod(m) if the inverse exists. In fact, this property of the algorithm is probably the most often used in public key cryptography.

To find the inverse of element e modulo m, set $r_0 = m$ and $r_1 = e$. Perform the algorithm sketched in Table 19-1. When r_{j+2} becomes zero, examine r_{j+1}.

Table 19-1

The flow of the extended Euclidean algorithm

j	r_j	r_{j+1}	d_{j+1}	r_{j+2}	s_j	t_j
0	341	231	1	110	1	0
1	231	110	2	11	0	1
2	110	11	10	0	1	-1
3					-2	3

If it is unity, then e has an inverse modulo m and it is t_{j+1}. (For convenience, when calculating the inverse, you may want to perform the calculations for the $\{t_i\}$ in $\text{mod}(m)$).

Exercise 19

This exercise asks you to investigate whether or not an inverse exists to a particular element given a particular modulus.

1. Determine whether an inverse to 3 mod(101) exists. If so, find the inverse.

2. Determine whether an inverse to 41 mod(167) exists. If so, find the inverse.

Primitive Roots

Primitive roots are an important element in number theory. They are, in a very real sense, analogous to the primitive polynomials we studied a few modules back. Indeed, primitive roots also enable us to develop a long cycle of states.

The Euler-Fermat theorem guarantees us that $a^{\phi(m)} \equiv 1 \mod(m)$ if $(a, m) = 1$. This raises a related question. Is $\phi(m)$ the smallest positive exponent, d, of a such that $a^d \equiv 1 \mod(m)$? If it is, then d is called a primitive root of m. Let's construct the set

$$\{a^d \mod(m)\}, d = 1, 2, \ldots, \phi(m) \tag{20.1}$$

for two values of a for which $(a, m) = 1$.

If $a = 3$ and $m = 7$, we find that the set in Equation 20.1 is

$$\{3, 2, 6, 4, 5, 1\} \tag{20.2}$$

Note that $d = 6$; therefore, 3 is a primitive root of 7. If, however, we select $a = 2$, we get the following set:

$$\{2, 4, 1, 2, 4, 1\} \tag{20.3}$$

We see that for this case $d = 3$; therefore, 2 is not a primitive root of 7.

Incidentally, it is not hard to see that if a is a primitive root of m, then

$$\{a, a^2, \cdots, a^{\phi(m)}\} \tag{20.4}$$

constitutes a reduced residue set, Q, of m.

Not all numbers have primitive roots. They are possessed only by those m of the form

$$m = 1, 2, 4, p^n, 2p^n \tag{20.5}$$

where p is an odd prime.[1]

---

[1]For the special case $m = 2^n, n > 2$, if $a \equiv 3$ or $5 \mod(8)$, then $d = 2^{n-2}$. This result is behind many of the multiplicative congruential generators in software packages.

Let's take a look at a larger prime and one of its primitive roots. $m = 101$ is a prime and $a = 3$ is one of its primitive roots.[2] Table 20-1 tabulates the set in Equation 20.1.

Table 20-1

The set in Equation 20.1 for $m = 101$ and $a = 3$

i	3^i mod(101)	i	3^i mod101	i	3^i mod(101)	i	3^i mod(101)	i	3^i mod(101)
1	3	21	50	41	59	61	7	81	83
2	9	22	49	42	76	62	21	82	47
3	27	23	46	43	26	63	63	83	40
4	81	24	37	44	78	64	88	84	19
5	41	25	10	45	32	65	62	85	57
6	22	26	30	46	96	66	85	86	70
7	66	27	90	47	86	67	53	87	8
8	97	28	68	48	56	68	58	88	24
9	89	29	2	49	67	69	73	89	72
10	65	30	6	50	100	70	17	90	14
11	94	31	18	51	98	71	51	91	42
12	80	32	54	52	92	72	52	92	25
13	38	33	61	53	74	73	55	93	75
14	13	34	82	54	20	74	64	94	23
15	39	35	44	55	60	75	91	95	69
16	16	36	31	56	79	76	71	96	5
17	48	37	93	57	35	77	11	97	15
18	43	38	77	58	4	78	33	98	45
19	28	39	29	59	12	79	99	99	34
20	84	40	87	60	36	80	95	100	1

[2]If a number, m, is of the form in Equation 20.5, then it possesses $\phi(\phi(m))$ primitive roots.

Recall that in Module 6, "Randomness I," we looked at a similar data set. We would expect to see a visible structure if we plotted $3^{i+1} \bmod(101)$ versus $3^i \bmod(101)$. However, if all we look at is a single instance, a single value of $3^i \bmod(101)$, we can see that we would have a tough problem identifying the i to which 3 has been raised and the exponentiation subsequently reduced modulo 101. This is at the heart of the first public key system we will look at. This is an example of a process that is relatively easy in one direction and relatively hard in the other. It is easy to find $3^i \bmod(101)$ given i, but hard to find i given $3^i \bmod(101)$. The easy problem involves exponentiation with subsequent modular reduction. The hard problem has a famous descriptive name: the *discrete logarithm problem*.

Exercise 20

In this exercise, all we're going to do is to verify two entries in Table 20-1. We'll use these two specific entries in Module 23.

1. Find $3^{70} \bmod(101)$ and $3^{87} \bmod(101)$.

Chinese Remainder Theorem

The Chinese Remainder Theorem is a useful technique for working with a mixed-radix number system. Although we are used to uniradix systems, such as counting in base 10 or base 2, we use mixed-radix systems all the time, for example, with latitude/longitude and, of course, time. We need the Chinese Remainder Theorem for Module 29, "Secret Sharing." The material may be of help to you for all sorts of other things too.

The Chinese Remainder Theorem enables us to find an answer to the following special set of simultaneous equations:

$$x \equiv r_1 \, \mathrm{mod}(m_1) \qquad\qquad (21.1)$$

$$x \equiv r_2 \, \mathrm{mod}(m_2)$$

$$\vdots$$

$$x \equiv r_n \, \mathrm{mod}(m_n)$$

If all the moduli of Equation 21.1 are relatively prime, that is, $(m_i, m_j) = 1$ for all i and j, $i \neq j$, then x will be unique modulo M where

$$M = m_1 m_2 \ldots m_n \qquad\qquad (21.2)$$

The solution can be written in a form ready for evaluation by first defining two sets of auxiliary variables. The first of these is the set $\{M_j\}$ whose members are defined as

$$M_j = \frac{M}{m_j} \qquad\qquad (21.3)$$

The second set of auxiliary variables is the set $\{\rho_j\}$ where the members are computed according to

$$\rho_j M_j \equiv 1 \mathrm{mod}(m_j) \qquad\qquad (21.4)$$

The solution to Equation 21.1 is

$$x \equiv (\rho_1 r_1 M_1 + \rho_2 r_2 M_2 + \cdots + \rho_n r_n M_n) \mathrm{mod}(M) \qquad\qquad (21.5)$$

▬ ▬ Exercise 21

This exercise asks you to formulate a mixed-radix problem in the form of Equation 21.1 and then to compute its solution using Equation 21.5.

1. Imagine that you have three counters commonly counting a train of pulses as depicted in Figure 21-1. Counter i is started at zero and counts up to $c_i - 1$ and then resets to zero at the next count. The count of counter i is displayed as r_i.

   Let $(c_1, c_2, c_3) = (3, 5, 7)$. If at the end of counting pulses, the counters' counts are $(r_1, r_2, r_3) = (1, 0, 6)$, what is the minimum number of pulses that were counted?

Figure 21-1
Three counters
counting pulses

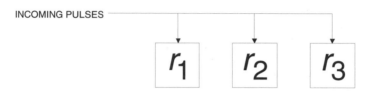

INCOMING PULSES

PART III

Introduction to Public Key Cryptography

The third part of this book comprises six modules that provide an introduction to the relatively new scientific area of cryptography. Public key cryptography, or asymmetric cryptography, is fascinating because it is counterintuitive. The idea that two parties, sharing nothing in secret a priori, can develop a mutually held secret quantity based on a publicly viewable message exchange sounds outrageous at first. It also sounds as though it may be extremely useful; indeed, it is.

It is dangerous to suspect that public key cryptography behaves like symmetric cryptography. The two are quite different in almost all aspects. The two share no similarities, and public key cryptography has all sorts of special traps and snares. We do not spend much time studying these traps; they are best covered in a more advanced work, particularly one that deals with cryptography protocols.

I do, however, expect that you will gain an understanding, appreciation, and basic working knowledge of public key cryptography from studying and reflecting on the material contained in this part.

Merkle's Puzzle

Merkle's Puzzle was an extremely important milestone in the development of public key cryptography. It was important not because it presented a highly effective or efficient system as published, but because it showed that a passive eavesdropper could be required to expend a considerably greater amount of computation than two correspondents establishing a secret quantity between each other with no a priori secret knowledge as required by a one-key cryptographic system. This result provided hope and engendered enthusiasm that public key cryptography would eventually become a real and viable tool for the cryptographic sciences.

As originally presented, Merkle's Puzzle involves the creation and publication of P puzzles.[1] A puzzle is a cryptogram that is solved by exhaustion —that is, all of the possible keying variables are posited until the cryptogram is broken. A puzzle has a particular form. It consists of three fields of fixed length, as depicted in Figure 22-1.

The three fields and their functions are

- **Puzzle Number**   A unique number for each puzzle
- **Keying Variable**   The keying variable to be used by the correspondents in a one-key cryptosystem such as the *Data Encryption Standard* (DES)
- **Known Text**   A string of characters that is the same for all puzzles and is published for all to know

Each puzzle is formed by encrypting all three fields of the puzzle using a known one-key cryptosystem with a unique (to each puzzle) keying variable chosen by the puzzle's creator. The key aspect of the system is that the keying variable chosen to encrypt a particular puzzle is chosen from a set of keying variables that is large but not so large that they cannot all be tried within a reasonable time; in other words, solving a particular puzzle is costly but doable.

The system functions as follows:

1. Party A creates, stores, and publishes P puzzles. The puzzles are published in a random order.

Figure 22-1
The form of
the puzzle

PUZZLE NUMBER | KEYING VARIABLE | KNOWN TEXT

---

[1] R. Merkle, "Secure Communications over Insecure Channels," *Communications of the ACM* 21 (1978): 294–9.

2. Party B selects one of the published puzzles randomly.

3. Party B solves the puzzle it selected by exhausting all the possible keying variables. Party B knows when it has solved the puzzle because the Known Text appears upon decryption using the correct keying variable.

4. Party B notes the Puzzle Number that is also disclosed upon successful decryption.

5. Party B publicly informs Party A of the Puzzle Number that it has solved.

6. Parties A and B now use the Keying Variable (the second field in Figure 22-1, which is disclosed upon successful decryption of the puzzle) as a mutually held cryptovariable and commence one-key-protected cryptographic communications between them using a system such as the DES.

If there are V possible keying variables, and if it requires a certain amount of work (W) to either decrypt a puzzle, using a correctly posited keying variable, or reject a posited keying variable, then the average work required to decrypt a puzzle is $\frac{V}{2} \times W$. This is the average amount of work that it will take Party B to establish a working keying variable with Party A.

Now, in order to find the keying variable in use between Parties A and B, an eavesdropper must, on average, expend an amount of work $\frac{P}{2} \times \frac{V}{2} \times W$.

The eavesdropper must therefore expend $\frac{P}{2}$ as much work, on average, as Party B. This leverage is an essential characteristic of Merkle's Puzzle.

Merkle's Puzzle may be implemented by selecting a high-grade one-key cryptographic system such as a block cipher function operated in a confidentiality mode and restricting the keying variable to $K_0 < K$ bits. For example, let's assume that we have picked a block cipher function whose keying variable, K, is 128 bits. If we fix 108 of those bits and publicly announce what they are, then $V = 2^{20}$. Party B only needs to exhaust those remaining 20 bits to solve a particular puzzle. On average, therefore, Party B would need to make 2^{19} attempts to solve the puzzle.

If P is chosen so that $P \cong V$, then the eavesdropper is forced into an amount of work on the order of the square of the amount of work required of Party B.

Exercise 22

This exercise asks you to consider one of the important points for making Merkle's Puzzle work.

1. In Merkle's Puzzle, what would be the consequence if the puzzle-generating party does not send the puzzles out in a random order, but sends them sequentially by puzzle number?

23

The Diffie-Hellman Public Key Cryptographic System

The Diffie-Hellman public key cryptographic system[1,2] was the first published public key cryptographic system. Its security depends on the existence of a process that is relatively easy to perform in one direction and relatively hard to perform in another. Like many public key cryptographic systems, the Diffie-Hellman system is also subject to an active cryptographic attack that has earned the name man-in-the-middle attack. *We'll take a look at this vulnerability.*

In Module 20, "Primitive Roots," we talked about the discrete logarithm problem and how it seems to be a much harder problem than its inverse problem, exponentiation. The Diffie-Hellman system makes use of this apparent asymmetric complexity to create a public key cryptographic system. Such a system enables a secret quantity to be developed by two different parties who have no prior private agreements. They can develop their mutually held secret quantity in open communications.

The implementation is quite straightforward. It starts with two parties: Party A and Party B. Both parties decide (publicly) on a prime number and a primitive root of that prime number. Let's call the prime number p and the primitive root a.

Once the prime and primitive root have been agreed upon, the following process takes place:

■ Party A generates an exponent, A, in the range of $1 < A < p$.

■ Party A computes $a^A \bmod(p)$ and sends the result to Party B.

■ Party B generates an exponent, B, in the range of $1 < B < p$.

■ Party B computes $a^B \bmod(p)$ and sends the result to Party A.

■ Party B receives $a^A \bmod(p)$, raises it to the B power, and reduces the result $\bmod(p)$. Thus, Party B has computed

$$(a^A)^B \bmod(p) \tag{23.1}$$

■ Party A receives $a^B \bmod(p)$, raises it to the A power, and reduces the result $\bmod(p)$. Thus, Party A has computed

$$(a^B)^A \bmod(p) \tag{23.2}$$

---

[1]W. Diffie and M. Hellman, "New Directions in Cryptography," *IEEE Transactions on Information Theory* IT-22, no. 6 (1976): 644–54

[2]W. Diffie and M. Hellman, "Privacy and Authentication: An Introduction to Cryptography," *Proceedings of the IEEE* 67 (1979): 397–427.

Note that Equations 23.1 and 23.2 are the same quantity:

$$a^{AB} \bmod(p) \tag{23.3}$$

This quantity has been developed by both Parties A and B using only the exponentiation operation. In order for another party, say, Party C, to derive Equation 23.3, it would have to solve the discrete logarithm problem and recover either A from $a^A \bmod(p)$ or B from $a^B \bmod(p)$.

A public key cryptographic system is often referred to as an *asymmetric cryptographic system* or a *two-key cryptographic system* because it involves the generation of two different secret keying variables: one by Party A and one by Party B. Note that A generated by Party A and B generated by Party B are never revealed to anyone and are never known by the other party.

Once both parties have developed the mutually held secret quantity $a^{AB} \bmod(p)$, they can use it for keytext or as a secret keying variable for a one-key cryptographic system such as the one used in the paradigm of Figure 4-1 of Module 4, "Toward a Cryptographic Paradigm."

Let's look at an example. Suppose Parties A and B decide upon the prime $p = 101$ and the primitive root $a = 3$. Let's also suppose that Party A picks $A = 5$ and Party B picks $B = 6$. Using the results displayed in Table 20-1 of Module 20, we see that

$$3^5 \bmod(101) \equiv 41 \tag{23.4}$$

and

$$3^6 \bmod(101) \equiv 22 \tag{23.5}$$

Party B must compute $(41)^6 \bmod(101)$. The procedure is simple. If you square 41, Party B gets 1,681, which leaves a remainder of 65 when reduced $\bmod(101)$. Party B now knows that 65 is 41 squared $\bmod(101)$. If you square 65, Party B gets 4,225, which leaves a remainder of 84 when reduced $\bmod(101)$. Party B now has 41 to the fourth power $\bmod(101)$. So, Party B finally multiplies 65 by 84 and gets 5,460, which leaves a remainder of 6 when reduced $\bmod(101)$. This is 41 to the sixth power $\bmod(101)$.

Party A must compute $(22)^5 \bmod(101)$. This can be done in many ways. If you square 22, Party A gets 484, which leaves a remainder of 80 when reduced $\bmod(101)$. Party A now knows that 80 is 22 squared $\bmod(101)$. Party A now squares 80, getting 6,400. If you divide by 101, Party A obtains a remainder of 37 so Party A knows that 22 to the fourth power is 37 $\bmod(101)$. Finally, Party A multiplies 37 by 22, gets 814, and divides by 101

Figure 23-1
Man-in-the-middle
attack[3]

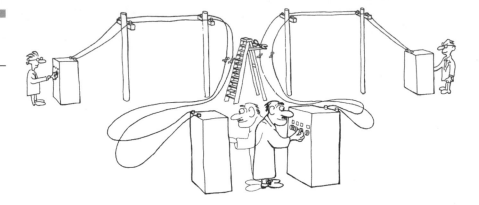

to get a remainder of 6, which is the fifth power of 22 mod(101) and also the same number that Party B computed, as expected.

A type of active attack can be wrought against the Diffie-Hellman system. It is sometimes called the *active transparency attack*, but is more commonly known as the *man-in-the-middle attack*. As illustrated in Figure 23-1, if Party C interposes itself between Parties A and B, it can deceive, or *spoof*, Parties A and B into believing that they are communicating with each other, thereby compromising the secret communications they attempt to establish.

The attack functions as follows:

1. Party A computes a^A mod(p) and sends it to Party B.
2. Party B does not receive a^A; Party C receives it instead.
3. Party C computes a^C mod(p) and sends it to Party A.
4. Party C also sends a^C to Party B.
5. Party C computes $(a^A)^C = a^{AC}$.
6. Party A computes $(a^C)^A = a^{AC}$.
7. Party B computes a^B mod(p) and sends it to Party A.
8. Party A does not receive a^B; Party C receives it instead.
9. Party C computes $(a^B)^C = a^{BC}$.
10. Party B computes $(a^C)^B = a^{BC}$.

---

[3]Diagram appropriated from "The Discrete Logarithm Public Cryptographic System," *NTIA Report* 81–81, September 1981.

Parties A and C now hold a common secret. Parties B and C also hold a common secret. However, Parties A and C believe that they hold the same common secret. Party C can now exploit any secret communications that Parties A and B may initiate based on their supposed common secret.

Exercise 23

This exercise merely takes you through the Diffie-Hellman system once again. The concepts are not difficult, but the practice is worthwhile.

1. Parties A and B decide upon the prime $p = 101$ and the primitive root $a = 3$. Suppose that Party A picks $A = 70$ and Party B picks $B = 87$. What quantity is developed in common by Parties A and B? By direct computation, show that they both do indeed arrive at this value. You may use the results displayed in Table 20-1 of Module 20 and the result of the exercise in the module on the binary exponentiation algorithm to get started.

As a final note, it is hard to find primitive roots of large primes. B. Schneier[4] suggests that a primitive root is not really necessary. It just needs to generate a very large group. Also, S. Pohlig and M. Hellman[5] caution that $p - 1$ should not have only small prime factors. Ideally, $(p - 1)/2$ should also be a prime.

---

[4]B. Schneier, *Applied Cryptography*, 2nd edition, (John Wiley and Sons, Inc., 1996).

[5]S. Pohlig and M. Hellman, "An Improved Algorithm for Computing Logarithms over GF(p) and Its Cryptographic Significance," *IEEE Transactions on Information Theory*, IT-24, no. 1 (1978): 106–10.

Split Search

*In the previous module, we looked at the Diffie-Hellman public key crypto-
graphic system. We noted that its security depends on the existence of a
process that is relatively easy to perform in one direction and relatively hard
to perform in another. In this section, we examine a shortcut for solving the
Diffie-Hellman system. This method does not require complete exhaustion.*

In this short book, we do not pay much attention to cryptanalysis other
than in relation to the systematic exhaustion of the cryptovariables.
Exhaustion is, of course, what a cryptographer hopes will be required to
break his or her system, but quite often an attack is developed that
requires less effort than total exhaustion. When the effort becomes significantly less than exhaustion, the security of the cryptosystem may be
imperiled.

The Diffie-Hellman system's security depends on the difficulty of finding
the discrete logarithm. It is believed that this is a very difficult problem and
that modularly reduced exponentiation is relatively easy, but it is a one-way
function in that its inverse, the logarithm, is relatively hard.

A shortcut[1] for finding the logarithm is available that requires far less
effort than exhaustion, but the amount of work required for very large numbers still remains so extensive that the system's security is not in question.
It is worth our time to look at this as an example of a less than exhaustive
attack and also to gain a bit more familiarity with number theory.

The shortcut hinges on the realization that for any prime, p, the number,
q, where

$$q = m_1 \left\lceil \sqrt{p} \right\rceil + m_2 \tag{24.1}$$

and where $\lceil \cdot \rceil$ is the ceiling function defined as

$$\lceil x \rceil = \left\{ \begin{matrix} x, & x \in \{integers\} \\ \lfloor x \rfloor + 1, & x \notin \{integers\} \end{matrix} \right\} \tag{24.2}$$

will span all of the members of the principal complete residue set mod(p) if
the multipliers m_1 and m_2 are allowed to independently span the range

$$0 \le m_1, m_2 < \left\lceil \sqrt{p} \right\rceil$$

---

[1]D. Knuth, *The Art of Computer Programming*, vol. 1, 2nd. ed., (Reading, MA: Addison-
Wesley, 1973).

Our goal is to solve for q such that

$$r \equiv a^q \, mod(p) \tag{24.3}$$

or, equivalently,

$$r \equiv a^{m_1 \lceil \sqrt{p} \rceil + m_2} \, mod(p) \tag{24.4}$$

Let's rewrite Equation 24.4 as

$$ra^{-m_2} \equiv a^{m_1 \lceil \sqrt{p} \rceil} \, mod(p) \tag{24.5}$$

For our upcoming computations, let's calculate a^{-m_2} by calculating $(a^{-1})^{m_2}$.

The next step is to create two tables: one for the left side of Equation 24.5 and the other for the right side of the equation. Two entries of these tables must be the same. This match enables us to determine q.

For our example, let's use the problem statement from the exercise in Module 20, "Primitive Roots," for which $p = 101$ and $3^q \equiv 17$ mod(101). For this case, $\left\lceil \sqrt{101} \right\rceil = 11$ and 3^{-1} mod(101) $\equiv 34$. We form the two tables shown in Table 24-1.

Table 24-1

The two tables for performing the split search

m_1	$3^{11 m_1}$ mod(101)	17×34^{m_2} mod(101)	m_2
0	1	17	0
1	94	73	1
2	49	58	2
3	61	53	3
4	78	**85**	4
5	60	62	5
6	**85**	88	6
7	11	63	7
8	24	21	8
9	34	7	9
10	65	36	10

Note that there is indeed an entry on the left that is the same as the entry on the right, namely 85. The entry on the left corresponds to $m_1 = 6$ and the entry on the right corresponds to $m_2 = 4$. From this information, we can quickly ascertain q from Equation 24.1:

$$q = 6 \times 11 + 4$$

$$= 70 \qquad\qquad (24.6)$$

Exercise 24

This exercise goes through the split search for one more example. The example chosen is that of Party B's choice for a secret exponent in the exercise in Module 23, "The Diffie-Hellman Public Key Cryptographic System."

1. For our example, let's use the problem statement from the exercise in Module 23 for which $p = 101$ and $3^q \equiv 8 \bmod(101)$.

A Variant of the Diffie-Hellman System

The Diffie-Hellman system has an important variant. It is based on primitive polynomials instead of primitive roots. It can be run very quickly, especially in hardware, as the computations are quite naturally implementable in base-2 arithmetic.

In its pristine form, the Diffie-Hellman system relies on using powers of a primitive root modulo a prime claiming the primitive root. This has a useful counterpart, which was discovered in a paper by S. Berkovits et al.[1] This latter scheme operates in precisely the same manner except that instead of exchanging integers, Parties A and B exchange polynomials and instead of reducing modulo a prime, reduction is performed modulo a primitive polynomial.

We need a little background before we plunge in. This background is simple and will seem very natural as you are already familiar with the operations and concepts.

First, we consider a general polynomial of degree $n - 1$ of the form

$$c_{n-1}x^{n-1} + \cdots + c_1 x + c_0 \tag{25.1}$$

where each of the coefficients $\{c_i\}$ is either a one or zero. We then define the addition of two degree $n - 1$ polynomials, $a_{n-1}x^{n-1} + \cdots + a_1 x + a_0$ and $b_{n-1}x^{n-1} + \cdots + b_1 x + b_0$, as a degree $n - 1$ polynomial $r_{n-1}x^{n-1} + \cdots + r_1 x + r_0$, where r_i is the modulo two sum of a_i and b_i.

There are 2^n polynomials of the form in Equation 25.1 and these polynomials form an Abelian (commutative) group under addition as we just defined it.

We can also define the multiplication of polynomials. We cannot use ordinary polynomial multiplication if we want our results to stay within the form in Equation 25.1 because the product in ordinary polynomial multiplication yields a polynomial of a degree greater than $n - 1$. To see this, simply multiply two degree $n - 1$ polynomials together:

$$
\begin{aligned}
&(a_{n-1}x^{n-1} + \cdots + a_1 x + a_0)(b_{n-1}x^{n-1} + \cdots + b_1 x + b_0) \\
&= a_{n-1}b_{n-1}x^{2n-2} + (a_{n-1}b_{n-2} + a_{n-2}b_{n-1})x^{2n-3} + \cdots + a_0 b_0
\end{aligned} \tag{25.2}
$$

---

[1]S. Berkovits, J. Kowalchuk, and B. Schanning, "Implementing Public Key Scheme," *IEEE Communications Magazine* 17, no. 3 (May 1979).

Therefore, we will have to manage it so that polynomials of, or exceeding, the n-th degree will be somehow reduced or mapped into a polynomial of degree $n - 1$ or less. Such a mapping can be done by modular reduction using a primitive polynomial.

Remember when we did modular reduction using integers? Let's say we reduced 11 modulo 8 to get the result into 8's principal complete residue set. We do this by simply subtracting or adding the modulus as often as we want. In effect, the modulus was equivalent to a zero and indeed the modulus is congruent to zero. The same may be done with a polynomial.

As with integers, polynomials may be factored. The polynomial $x^2 + 1$ cannot be factored if the polynomials have term coefficients from the set of real numbers because two factors of the form

$$(x + a)(x + b) = x^2 + 1 \tag{25.3}$$

cannot be found. However, if we operate with polynomials whose term coefficients are zeros and ones and we reduce modulo two, then $x^2 + 1$ may be factored. Indeed, it is a perfect square

$$(x + 1)(x + 1) \equiv x^2 + 1 \tag{25.4}$$

where the congruence is understood to be modulo two; therefore, the term $2x$ that results in squaring $x + 1$ is congruent to zero.

If a polynomial can be factored into a product of polynomials of a smaller degree, then the polynomial is said to be *reducible*. Polynomials that cannot be factored this way arc termed *irreducible*. This is analogous to composite and prime numbers, but this analogy ends with a further dichotomization of irreducible polynomials. This further partition breaks irreducible polynomials into primitive polynomials and irreducible non-primitive polynomials.[2]

Primitive polynomials were introduced in Module 8, "m-sequences." They have remarkable properties and were used to specify m-sequence generators. In our case, they may be used as a modulus of reduction so that the set

$$\{0, a, a^2, \cdots, a^{2^n - 1}\} \tag{25.5}$$

---

[2] A primitive polynomial is irreducible. An irreducible polynomial is not primitive in general. It is primitive if $2^n - 1$ is a prime, however. A prime of this form is called a *Mersenne prime*.

where a represents a primitive element, maps one-to-one onto the 2^n polynomials of the form in Equation 25.1.

For the primitive polynomials listed in this book, you may consider that $a = x$ is a primitive element. We know from Table 8-2 in Module 8 that $x^4 + x + 1$ is a primitive polynomial. Let's see how we can use it to generate the nonzero members of 25.5. Figure 25-1 shows successive exponentiations of $a = x$ modulo $x^4 + x + 1$. Note that we are considering the primitive polynomial as equivalent to zero. In other words,

$$x^4 + x + 1 = 0 \qquad\qquad (25.6)$$

Figure 25-1
Exponentiation of
$a = x$ modulo
$x^4 + x + 1$
_____

$x^1 = x$

$x^2 = x^2$

$x^3 = x^3$

$x^4 = x^4$
$\quad = x+1^*$

$x^5 = x^2 + x$

$x^6 = x^3 + x^2$

$x^7 = x^4 + x^3$
$\quad = x^3 + x+1^*$

$x^8 = x^4 + x^2 + x$
$\quad = x^2 +1^*$

$x^9 = x^3 + x$

$x^{10} = x^4 + x^2$
$\quad = x^2 + x+1^*$

$x^{11} = x^3 + x^2 + x$

$x^{12} = x^4 + x^3 + x^2$
$\quad = x^3 + x^2 + x+1^*$

$x^{13} = x^4 + x^3 + x^2 + x$
$\quad = x^3 + x^2 +1^*$

$x^{14} = x^4 + x^3 + x$
$\quad = x^3 +1^*$

$x^{15} = x^4 + x$
$\quad =1^*$

(* Denotes result of modular reduction)

We can rewrite Equation 25.6 as

$$x^4 = x + 1 \tag{25.7}$$

by adding (modulo two) $x + 1$ to both sides of Equation 25.6. Thus, in Figure 25-1, when we raised x to the fourth power, we reduced it to $x + 1$.

This variant of the Diffie-Hellman system is performed in the same way as with integers. The variant system has the feature that the secretly developed quantity between Parties A and B is a polynomial and that polynomial may be considered as an n-bit quantity. For example, if $n = 4$, then the polynomial that results after modular reduction is of the form

$$c_3 x^3 + c_2 x^2 + c_1 x + c_0 \tag{25.8}$$

where the $\{c_i\}$ are zeros and ones. Parties A and B can simply use

$$(c_3, c_2, c_1, c_0) \tag{25.9}$$

as a secret 4-bit vector. Also, when Parties A and B are exchanging polynomials, they only need to send n-bit vectors to each other.

When using the system, exponentiation may be performed using the binary exponentiation algorithm just as was done with integers. Simple binary logic circuits can be straightforwardly constructed to perform all of the operations for this variant system and the system can be made to run extremely fast.

Let's look at an example using the primitive polynomial $x^7 + x + 1$. Table 25-1 shows the primitive element x to the powers 1, 2, 3, ..., 126 reduced modulo $x^7 + x + 1$. Let Party A choose $A = 56$ and Party B choose $B = 18$.

Party A computes

$$x^{56} \bmod (x^7 + x + 1) \equiv x^2 + x + 1 \tag{25.10}$$

and sends 0000111 (the 7 bits representing $x^2 + x + 1$) to Party B. Meanwhile, Party B has computed

$$x^{18} \bmod (x^7 + x + 1) \equiv x^6 + x^4 \tag{25.11}$$

and sends 1010000 to Party A.

Party A computes

$$(x^6 + x^4)^{56} \bmod(x^7 + x + 1) \equiv x^5 + x^4 + x^3 + x^2 + x + 1 = 0111111 \qquad (25.12)$$

Party B computes

$$(x^2 + x + 1)^{18} \bmod(x^7 + x + 1) \equiv x^5 + x^4 + x^3 + x^2 + x + 1 = 0111111 \qquad (25.13)$$

Both parties now hold $x^5 + x^4 + x^3 + x^2 + x + 1$ or 0111111, which is known only to them.

Architecturally, you may want to build this system on a large Mersenne prime. The following are three primitive trinomials corresponding to three large consecutive Mersenne primes:[3]

$$x^{521} + x^{32} + 1, 2^{521} - 1 \text{ is prime.}$$

$$x^{607} + x^{105} + 1, 2^{607} - 1 \text{ is prime.}$$

$$x^{1279} + x^{216} + 1, 2^{1279} - 1 \text{ is prime.} \qquad (25.14)$$

Table 25-1

Exponentiation of
a = x modulo
x^7 + x + 1

i	$x^i \bmod(x^7 + x + 1)$	i	$x^i \bmod(x^7 + x + 1)$
1	x	9	$x^3 + x^2$
2	x^2	10	$x^4 + x^3$
3	x^3	11	$x^5 + x^4$
4	x^4	12	$x^6 + x^5$
5	x^5	13	$x^6 + x + 1$
6	x^6	14	$x^2 + 1$
7	$x + 1$	15	$x^3 + x$
8	$x^2 + x$	16	$x^4 + x^2$

[3]N. Zierler, "Primitive Trinomials whose Degree Is a Mersenne Exponent," *Information and Control* 15 (1969): 67–69.

Table 25-1

Exponentiation of
a = x modulo
$x^7 + x + 1$
(continued)

i	$x^i \bmod(x^7 + x + 1)$	i	$x^i \bmod(x^7 + x + 1)$
17	$x^5 + x^3$	46	$x^6 + x^3 + x^2 + x$
18	$x^6 + x^4$	47	$x^4 + x^3 + x^2 + x + 1$
19	$x^5 + x + 1$	48	$x^5 + x^4 + x^3 + x^2 + x$
20	$x^6 + x^2 + x$	49	$x^6 + x^5 + x^4 + x^3 + x^2$
21	$x^3 + x^2 + x + 1$	50	$x^6 + x^5 + x^4 + x^3 + x + 1$
22	$x^4 + x^3 + x^2 + x$	51	$x^6 + x^5 + x^4 + x^2 + 1$
23	$x^5 + x^4 + x^3 + x^2$	52	$x^6 + x^5 + x^3 + 1$
24	$x^6 + x^5 + x^4 + x^3$	53	$x^6 + x^4 + 1$
25	$x^6 + x^5 + x^4 + x + 1$	54	$x^5 + 1$
26	$x^6 + x^5 + x^2 + 1$	55	$x^6 + x$
27	$x^6 + x^3 + 1$	56	$x^2 + x + 1$
28	$x^4 + 1$	57	$x^3 + x^2 + x$
29	$x^5 + x$	58	$x^4 + x^3 + x^2$
30	$x^6 + x^2$	59	$x^5 + x^4 + x^3$
31	$x^3 + x + 1$	60	$x^6 + x^5 + x^4$
32	$x^4 + x^2 + x$	61	$x^6 + x^5 + x + 1$
33	$x^5 + x^3 + x^2$	62	$x^6 + x^2 + 1$
34	$x^6 + x^4 + x^3$	63	$x^3 + 1$
35	$x^5 + x^4 + x + 1$	64	$x^4 + x$
36	$x^6 + x^5 + x^2 + x$	65	$x^5 + x^2$
37	$x^6 + x^3 + x^2 + x + 1$	66	$x^6 + x^3$
38	$x^4 + x^3 + x^2 + 1$	67	$x^4 + x + 1$
39	$x^5 + x^4 + x^3 + x$	68	$x^5 + x^2 + x$
40	$x^6 + x^5 + x^4 + x^2$	69	$x^6 + x^3 + x^2$
41	$x^6 + x^5 + x^3 + x + 1$	70	$x^4 + x^3 + x + 1$
42	$x^6 + x^4 + x^2 + 1$	71	$x^5 + x^4 + x^2 + x$
43	$x^5 + x^3 + 1$	72	$x^6 + x^5 + x^3 + x^2$
44	$x^6 + x^4 + x$	73	$x^6 + x^4 + x^3 + x + 1$
45	$x^5 + x^2 + x + 1$	74	$x^5 + x^4 + x^2 + 1$

(continued)

Table 25-1

Exponentiation of $a = x$ modulo $x^7 + x + 1$ (continued)

i	$x^i \bmod(x^7 + x + 1)$	i	$x^i \bmod(x^7 + x + 1)$
75	$x^6 + x^5 + x^3 + x$	103	$x^6 + x^4 + x^3 + x^2 + 1$
76	$x^6 + x^4 + x^2 + x + 1$	104	$x^5 + x^4 + x^3 + 1$
77	$x^5 + x^3 + x^2 + 1$	105	$x^6 + x^5 + x^4 + x$
78	$x^6 + x^4 + x^3 + x$	106	$x^6 + x^5 + x^2 + x + 1$
79	$x^5 + x^4 + x^2 + x + 1$	107	$x^6 + x^3 + x^2 + 1$
80	$x^6 + x^5 + x^3 + x^2 + x$	108	$x^4 + x^3 + 1$
81	$x^6 + x^4 + x^3 + x^2 + x + 1$	109	$x^5 + x^4 + x$
82	$x^5 + x^4 + x^3 + x^2 + 1$	110	$x^6 + x^5 + x^2$
83	$x^6 + x^5 + x^4 + x^3 + x$	111	$x^6 + x^3 + x + 1$
84	$x^6 + x^5 + x^4 + x^2 + x + 1$	112	$x^4 + x^2 + 1$
85	$x^6 + x^5 + x^3 + x^2 + 1$	113	$x^5 + x^3 + x$
86	$x^6 + x^4 + x^3 + 1$	114	$x^6 + x^4 + x^2$
87	$x^5 + x^4 + 1$	115	$x^5 + x^3 + x + 1$
88	$x^6 + x^5 + x$	116	$x^6 + x^4 + x^2$
89	$x^6 + x^2 + x + 1$	117	$x^5 + x^3 + x^2 + x + 1$
90	$x^3 + x^2 + 1$	118	$x^6 + x^4 + x^3 + x^2 + x$
91	$x^4 + x^3 + x$	119	$x^5 + x^4 + x^3 + x^2 + x + 1$
92	$x^5 + x^4 + x^2$	120	$x^6 + x^5 + x^4 + x^3 + x^2 + x$
93	$x^6 + x^5 + x^3$	121	$x^6 + x^5 + x^4 + x^3 + x^2 + x + 1$
94	$x^6 + x^4 + x + 1$	122	$x^6 + x^5 + x^4 + x^3 + x^2 + 1$
95	$x^5 + x^2 + 1$	123	$x^6 + x^5 + x^4 + x^3 + 1$
96	$x^6 + x^3 + x$	124	$x^6 + x^5 + x^4 + 1$
97	$x^4 + x^2 + x + 1$	125	$x^6 + x^5 + 1$
98	$x^5 + x^3 + x^2 + x$	126	$x^6 + 1$
99	$x^6 + x^4 + x^3 + x^2$		
100	$x^5 + x^4 + x^3 + x + 1$		
101	$x^6 + x^5 + x^4 + x^2 + x$		
102	$x^6 + x^5 + x^3 + x^2 + x + 1$		

Exercise 25

As we have said, the polynomial variant of the Diffie-Hellman system is almost the same as the Diffie-Hellman system except that it uses polynomials. In this problem, we take a look at the split-search attack on the polynomial version.

1. Find q such that

$$x^q \equiv (x^5 + x^3 + x + 1) \bmod (x^7 + x + 1) \tag{25.15}$$

The RSA Public Key Cryptographic System

The best known public key cryptosystem is known by the trigraph RSA. RSA stands for the last names of the individuals who first publicly published the algorithm: Rivest, Shamir, and Adleman. It is an extremely important system and much of the financial security infrastructure relies on it.

A user of the RSA system[1] provides two public quantities, which are generally known as n and e. They represent a modulus for reduction and an encryption exponent, respectively. A message, M, is encrypted and sent to the user who has posted the particular n and e. M is simply a number that lies within the interval $2 \leq M \leq n - 1$ and may be created in any number of ways. For example, if you were to restrict the plaintext message symbols to $A, B, C, \ldots, Y, Z$, space, you could simply assign values to the symbols, as shown in Table 26-1, and then write the message according to the symbol values. In doing so, we would have to preface the symbol values for A through J with a leading zero in order to avoid any ambiguity that might arise—for example, encoding BB as 11 rather than 0101, the former being interpretable as L. With this plan, the message THE becomes

$$190704 \hspace{4cm} (26.1)$$

Table 26-1

Plaintext symbols and their values

Symbol	Symbol Value	Symbol	Symbol Value	Symbol	Symbol Value
A	0	J	9	S	18
B	1	K	10	T	19
C	2	L	11	U	20
D	3	M	12	V	21
E	4	N	13	W	22
F	5	O	14	X	23
G	6	P	15	Y	24
H	7	Q	16	Z	25
I	8	R	17	space	26

[1]R. Rivest, A. Shamir, and L. Adleman, "A Method for Obtaining Digital Signatures and Public-Key Cryptosystems," *Communications of the ACM* 21 (1978): 120–6.

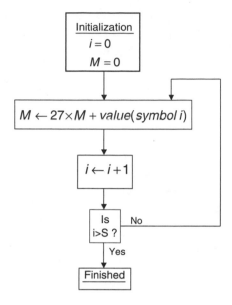

Figure 26-1
Radix-27 encoding

We could be a bit more efficient by using a radix-27 representation for encoding S symbols, as specified in Figure 26-1. With this encoding scheme, the message THE becomes $4 + 27 \times (7 + 27 \times 19)$ or

$$14044 \qquad (26.2)$$

which is considerably shorter (more efficient) than 26.1.

It is very simple to decode a radix-27 message. The last encoded symbol's value is the remainder of the message value divided by 27. The quotient is then divided by 27 and its remainder is the penultimate symbol's value, and so on.

So far, all we've done is prepare M for encryption in the RSA system for transmission to the party who has posted n and e. The next step is to compute

$$M^e \bmod(n) \qquad (26.3)$$

and send it.

The party who receives 26.3 and who has posted n and e decrypts 26.3 by raising this value to a secretly held number called the decryption exponent (d) and reducing $\bmod(n)$.

This is all very good, but how are e, d, and n created and how does this system work?

We will look first at the modulus (n). It is a special composite number. It is the product of two very large primes (p and q) that are of the same magnitude but not equal. We need the Euler phi (ϕ) function of n. This can be easily obtained:

$$\phi(n) = (p - 1)(q - 1) \tag{26.4}$$

Next, the preparer picks the encryption exponent (e) that will be publicly posted. The guidelines for picking e strongly suggest that e be a very large number and e and $\phi(n)$ be relatively prime.

Once e is selected, the preparer solves for d. The operative equation is

$$ed \equiv 1 \bmod(\phi(n)) \tag{26.5}$$

This can be solved by using the Euler-Fermat theorem or, more likely, the extended Euclidean algorithm. No matter how you do it, ensure that the chosen value of e also yields a very large value for d.

It should now be clear why the decryption process yields the original message M. It is because

$$(M^e)^d \equiv M \bmod(n) \tag{26.6}$$

Let's see an example. We pick

$$p = 11$$
$$q = 13 \tag{26.7}$$

Therefore,

$$n = 11 \times 13$$
$$= 143 \tag{26.8}$$

and we, arbitrarily, select

$$e = 7 \tag{26.9}$$

Now,

$$\phi(pq) = \phi(11 \times 13)$$
$$= \phi(11) \times \phi(13)$$
$$= 10 \times 12$$
$$= 120 \qquad (26.10)$$

We now need to find d such that

$$7d \equiv 1 \bmod(120) \qquad (26.11)$$

The solution of Equation 26.11 is easily determined:

$$d = 103 \qquad (26.12)$$

For our message, M, we'll select the very terse text:

$$M = 5 \qquad (26.13)$$

To encrypt M, we form

$$5^7 \bmod(143)$$
$$\equiv 47 \qquad (26.14)$$

Our encrypted message is thus 47. To decrypt it, we perform Equation 26.15:

$$47^{103} \bmod(143)$$
$$\equiv 5 \qquad (26.15)$$

and see that the method works.

At this point, we can make a few comments on the security of the RSA system. First, the RSA system has a characteristic that is not shared by symmetric cryptosystems. That characteristic is that a suspected plaintext can be tested merely by encrypting it. If the encrypted plaintext matches the ciphertext, then we have clearly determined the plaintext. A way to counter this technique of searching for the plaintext is to set aside part of M as a random data field to be discarded upon decryption. The search technique may be frustrated by generating the data for the random data field for each message sent.

Our second comment concerns the factoring problem. If n can be factored and p and q can be recovered, then $\phi(n)$ can be computed and d can be found. We can therefore assert that breaking the RSA system is no more difficult than a hard factoring problem. Unfortunately, this is only an upper bounding. We still do not know the shortest path to a solution.

Our third comment addresses yet another peculiar trait to this system that has no counterpart in symmetric cryptography. It has been given the name *cycling*. When we raised M^5 to the $d = 103$ power mod n in Equation 26.15, we recovered $M = 5$. We could have performed the exponentiation in many ways. The binary exponentiation algorithm is a prime candidate. However, let's take a long sequential approach and see if we can find something interesting. Table 26-2 computes $M^i \bmod (143)$ for $i = 1, 2, 3,$ $\ldots, 103$ for $M = 5$. Note that $M^i \bmod (143)$ has a cycle length of 20. In other words, for this case,

$$M^{i+20} \equiv M^i \bmod (143) \qquad (26.16)$$

We discover that the plaintext may show up with an exponent less than d. In other words, we might try to find the plaintext without attempting to factor n. We could merely keep forming $(M^i)^j \bmod (143)$ for the ever increasing and sequential j. We can be even more efficient. Because d must be relatively prime to $\phi(n)$, we do not need to consider the sequential values of j. We could simply raise M^i to the sequential powers of e^k, with $k = 1, 2, 3, \ldots$.

Fortunately, the cycling attack is not of great concern. It has been strongly suggested[2] that if p and q are randomly selected and are the same bit length, then the expected number of trials for the cycling algorithm to find M is at least $\sqrt[3]{p}$.

---

[2]A. Menezes, P. van Oorschot, and S. Vanstone, *Handbook of Applied Cryptography,* (CRC Press, 1997), which cites an unpublished 1991 manuscript by R. Rivest entitled "Are 'Strong' Primes Needed for RSA?"

Table 26-2

M^i mod(143) for
$i = 1, 2, 3, \ldots,$
103 for $M = 5$

i	M^i mod(143)	i	M^i mod(143)	i	M^i mod(143)	i	M^i mod(143)
1	47	27	31	53	60	79	70
2	64	28	27	54	103	80	1
3	5	29	125	55	122	81	47
4	92	30	12	56	14	82	64
5	34	31	135	57	86	83	5
6	25	32	53	58	38	84	92
7	31	33	60	59	70	85	34
8	27	34	103	60	1	86	25
9	125	35	122	61	47	87	31
10	12	36	14	62	64	88	27
11	135	37	86	63	5	89	125
12	53	38	38	64	92	90	12
13	60	39	70	65	34	91	135
14	103	40	1	66	25	92	53
15	122	41	47	67	31	93	60
16	14	42	64	68	27	94	103
17	86	43	5	69	125	95	122
18	38	44	92	70	12	96	14
19	70	45	34	71	135	97	86
20	1	46	25	72	53	98	38
21	47	47	31	73	60	99	70
22	64	48	27	74	103	100	1
23	5	49	125	75	122	101	47
24	92	50	12	76	14	102	64
25	34	51	135	77	86	103	5
26	25	52	53	78	38		

Our fourth comment concerns the use of the RSA in a special network. The problem we're about to encounter was discovered by Gustavus Simmons.[3] It is an example of one of the most difficult disciplines within cryptography: defining a secure protocol. This short book spends little time on the advanced topics of cryptanalysis and secure protocols, but we include this instance as an example of a protocol problem. It is especially intriguing because the problem arises through a series of seemingly reasonable steps. It is the combination of such steps that leads counterintuitively to the problem.

This scenario involves a large mobile data distribution service. It has been decided that the RSA cryptoalgorithm will be used and determined that the modular arithmetic associated with the RSA will be done in the hardware for increased speed. A circuit design engineer suggests that all units use the same modulus, n, in order to take advantage of certain modular circuits the engineer has developed. The wireless engineers and the security specialists agree that a common modulus will also reduce the cryptovariable distribution effort and so they agree to the plan. Sounds reasonable, doesn't it?

Now suppose that the same message, M, is sent to two different subscribers whose encryption keys are e_1 and e_2, respectively. Further assume that e_1 and e_2 are relatively prime. In this case, we have

$$M^{e_1} \equiv C_1 \mathrm{mod}(n) \tag{26.17}$$

$$M^{e_2} \equiv C_2 \mathrm{mod}(n)$$

where C_1 and C_2 are the two encrypted messages corresponding to e_1 and e_2, respectively.

Think back now to the extended Euclidean algorithm. Because $(e_1, e_2) = 1$, we can find integers s and t such that

$$se_1 + te_2 = 1 \tag{26.18}$$

Now, e_1 and e_2 are both positive; therefore, of the two coefficients, s and t, one will be positive and one will be negative. If s is the negative coefficient, we try to calculate C_1^{-1}. If t is the negative coefficient, we try to calculate C_2^{-1}. (Remember that the inverses, if they exist, may also be found via the extended Euclidean algorithm.)

---

[3]G. Simmons, "A 'Weak' Privacy Protocol Using the RSA Cryptoalgorithm," *Cryptologia* 7 (1983): 180–2.

If $(C_1, n) = 1$, then C_1^{-1} can be found. However, if C_1 and n are not relatively prime, then the extended Euclidean algorithm will remarkably produce one of the two prime factors of n—either p or q. If this happens, it not only compromises the message, but it also breaks the entire system.

In another case, suppose t is negative and we can compute C_2^{-1}. We can write

$$
\begin{aligned}
(C_1)^s (C_2^{-1})^{|t|} &\equiv (M^{e_1})^s ((M^{e_2})^{-1})^{|t|} \mathrm{mod}(n) \\
&\equiv M^{se_1 + te_2} \mathrm{mod}(n) \\
&\equiv M \, \mathrm{mod}(n)
\end{aligned}
\tag{26.19}
$$

Surprisingly, we have recovered M.

Finally, we'll say a word about authentication. It is possible to use the RSA cryptosystem to sign a message so that the signature cannot be repudiated later. Assume that Party A wants to send a message M to Party B so that Party B will know that the message must have come from Party A; that is, it was equivalently signed.

This can be done as follows,[4] where e_B is Party B's public encryption key, d_A is Party A's private decryption key, n_A is Party A's public modulus, and n_B is Party B's public modulus:

■ Party A computes $M^{d_A} \mathrm{mod}(n_A)$.

■ Party A then computes $(M^{d_A} \mathrm{mod}(n_A))^{e_B} \mathrm{mod}(n_B)$

■ Party B receives the previous result from Party A and applies its decryption key followed by Party A's public encryption key.

By the way, don't be tempted to encrypt before signing (that is, first computing $M^{e_B} \mathrm{mod}(n_B)$ and then raising the result to d_A and reducing $\mathrm{mod}(n_A)$. Under some operating conventions, this reversal in procedure, although perhaps appearing innocuous, may destroy the nonrepudiation.[5]

Finally, be sure you check the most recent literature on authentication before getting too serious about any one scheme. Some additional interesting caveats and advisories on the previous process can be found in Menezes et al.[6]

---

[4]See W. Diffie and M. Hellman, "Privacy and Authentication: An Introduction to Cryptography," *Proceedings of the IEEE* 67 (1979): 397–427.

[5]See B. Schneier, *Applied Cryptography*, 2nd edition, (New York: John Wiley and Sons, 1996).

[6]Menezes et al, op cit.

Exercise 26

These two exercises span the nuts and bolts for understanding the implications of a property of exponentials and what that property might mean for the RSA cryptosystem.

1. In our simple example of the RSA cryptosystem, we picked $e = 7$ in Equation 26.9. Could we just as reasonably have picked $e = 8$?

2. We know that

$$x^a \times x^b = x^{a+b} \tag{26.20}$$

What does $a^x \times b^x$ equal and what is the implication with respect to the RSA authentication procedure that we looked at?

Prime Numbers II

It seems that prime numbers are a sine qua non of much of public key cryptography. In Module 14, "Prime Numbers I," we learned that the density of primes becomes thinner as numbers grow larger, but many primes are still available. But how do we get a large prime number? The process is a bit of a roundabout, but is quite fascinating and eminently practical.

Remember the Euler-Fermat theorem? It proclaims that if $(a, m) = 1$, then $a^{\phi(m)} \equiv 1 \bmod(m)$. So, if m was a prime (p), then

$$a^{p-1} \equiv 1 \bmod(p) \tag{27.1}$$

We're immediately tempted by the converse to Equation 27.1. That is, if we find that

$$a^{m-1} \equiv 1 \bmod(m) \tag{27.2}$$

for an m relatively prime to a, can we conclude that m is prime? If we could conclude this, then we would have an efficient test for the primality of m. In other words, we could pick a large prime number candidate at random and see if Equation 27.2 works or not.

But the converse isn't true. For example,

$$2^{340} \equiv 1 \bmod(341) \tag{27.3}$$

Clearly, $(2,341) = 1$, but $341 = 11 \times 31$. When a composite number such as m satisfies Equation 27.2 for $(a, m) = 1$, we say that m is a *pseudoprime* to the base a.

Well, maybe all is not lost with this approach. Maybe if we checked each a for which $(a, m) = 1$, we would find at least one a for which Equation 27.2 would not hold if m was a composite number.

Again, we run into trouble. The trouble has a specific name: *Carmichael numbers*. A Carmichael number is a composite number, m, that satisfies Equation 27.2 for every a relatively prime to m. A Carmichael number is also referred to as an *absolute pseudoprime*. An infinite number of Carmichael numbers exist.[1] The three smallest are

---

[1]W. Alford, A. Granville, and C. Pomerance, "There Are Infinitely Many Carmichael Numbers," *Annals of Mathematics* 140 (1994): 703–22.

- $561 = 3 \times 11 \times 17$
- $1105 = 5 \times 13 \times 17$
- $1729 = 7 \times 13 \times 19$ (This number is special for another reason also.[2])

The community seems to have been driven to a probabilistic approach that involves an entity called a *strong pseudoprime*.

We approach it by first considering that a large candidate prime, m, can be written as

$$m - 1 = 2^s t \tag{27.4}$$

where t is odd. Now, if either of the two conditions

$$a^t \equiv 1 \bmod(m) \tag{27.5a}$$

$$a^{2^r t} \equiv (m - 1) \bmod(m), \text{ for some } 0 \le r < s \tag{27.5b}$$

holds and m is a composite number, then m is a strong pseudoprime. Also either Equation 27.5a or Equation 27.5b must hold if m is not composite but prime.

It turns out that we're on the verge of a successful test because it has been proven that an odd composite number cannot be a strong pseudoprime to all bases that are relatively prime to m. In fact (and this will shortly become a key point), no odd composite number, m, is a strong pseudoprime to more than one-fourth of the bases that are relatively prime to m.

Michael Rabin made good use of this fact with the following algorithm:[3]

- Step 1: Pick B different bases, $a_1, a_2, \ldots, a_B$.
- Step 2: See if m is relatively prime to each of the B bases.
- Step 3: See if m is a strong pseudoprime to each of the B bases.

If m passes Steps 2 and 3, then m is prime with a probability of $\ge 1 - 2^{-2B}$. This probability may be made insignificant with modestly large B.

---

[2] This number is also the smallest number that can be written in two ways as the sum of two cubes.

[3] M. Rabin, "Probabilistic Algorithm for Testing Primality," *Journal of Number Theory* 12 (1980): 128–38.

In practice, B. Schneier[4] advises that when generating an n-bit prime, start by setting both the high and low bits to ones, then generate the remaining bits from a balanced binary Bernoulli source, and then check to see if the candidate prime is indeed relatively prime to a few of the small primes $3, 5, 7, \ldots$.

---

[4] B. Schneier, *Applied Cryptography*, 2nd edition, (John Wiley and Sons, Inc., 1996).

Exercise 27

This exercise checks a number to see if it passes the test to be a prime or strong pseudoprime.

1. The number $m = 4033$ is a composite number as $4033 = 37 \times 109$.
   Check to see if m passes the test to be a prime or strong pseudoprime to bases 2 and 3.

Keying
Variables

This fourth part comprises four modules. These modules discuss and dissect some of the issues attending the cryptographic keying variable. It is the keying variable upon which rests the security of the cryptosystem. To be sure, we must use a good cryptographic principle but we assume that what we are using is known publicly and what has not been released, and what must be solved for by cryptanalysis, is the specific keying variable protecting a period of communications.

It is therefore of primary importance to generate, distribute, and deploy keying variables properly. This is not an easy or straightforward task in general. Indeed it may be extremely complicated. What is unusual is the oft lack of attention or resources committed to it that given its singular importance.

28

Keying
Variable
Distribution

The effective and secure distribution of keying variables for a symmetric cryptosystem is one of the most important, and often thankless, jobs of maintaining a cryptographic circuit or network. The combinatorics can make distribution and accounting a fierce, onerous, and downright awesome task. Over the years, various architectures have been considered. This module takes a look at this task and a few of these architectures.

We believe that the security of a cryptosystem does not rest in the cryptoalgorithm's secrecy because we expect that such knowledge will eventually be compromised or somehow be available to an interested party. For this reason, we have decided that the security of the cryptosystem resides in the secret cryptographic keying variable that is inserted into the cryptographic key generator before it processes traffic.

The effects of key distribution failure and keying variable management can be spectacular. One of the most famous cases of this occurred during the Battle of Tannenberg, a decisive German victory in World War I. Tannenberg, a town in East Prussia, was supposed to be invaded by two Russian armies in 1914. According to David Kahn,[1]

> . . . Russian communications were woefully inadequate.
>
> Their communications lay naked to the enemy. The general inefficiency that crippled the Russian mobilization had fouled up distribution of the new military cipher and its keys. . . . Before a fortnight had passed Russian signalmen were no longer even trying to encipher messages, but were passing them over the radio in the clear.
>
> Almost 100,000 Russians were taken prisoner. An estimated 30,000 were dead or missing. The Russian 2nd Army had ceased to exist.
>
> Hoffman, the architect of the victory, acknowledged its real cause. "We had an ally that I can only talk about after it is all over—we knew all the enemy's plans. The Russians sent out their wireless in clear." The case was clear-cut. Interception of unenciphered communications had awarded the Germans their triumph. Tannenberg, which gave Russia the first push on her long slide into ruin and revolution, was the first battle in the history of the world to be decided by cryptologic failure.

Let's posit some simple models of keying variable distribution and see what they imply. Let's start with four units that want to be able to communicate with each other. The units have communication links, as shown in Figure 28-1.

---

[1]David Kahn, *The Codebreakers*, (Macmillan, 1967).

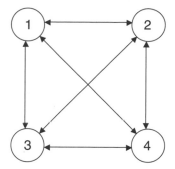

Figure 28-1
The four units and their communication links

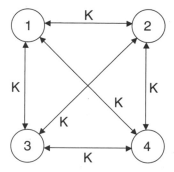

Figure 28-2
The four units all using the same keying variable

We may assign the same keying variable, K, to all of the links, as indicated in Figure 28-2. (The links can use the same keying variable with the *counter* [CTR] mode, for example, to communicate without incurring depth, as explained in Module 11, "Confidentiality Modes: Electronic Codebook [ECB] and Counter [CTR].")

We could also assign a different keying variable to all of the (six) separate links, as shown in Figure 28-3.

The assignment scheme of Figure 28-2 (assigning the same keying variable to all links) is highly efficient in that only one keying variable must be generated and loaded into the cryptosystem. It is also extremely hazardous to the overall security under a single compromise of that keying variable. One accident or loss, one mole in the system, one hostile acquisition destroys the entire network's security.

The assignment scheme of Figure 28-3, on the other hand, enables the network's security to degrade gracefully under compromise. However, the scheme of Figure 28-3 is extremely complex. It requires the production, distribution, maintenance, destruction, and accounting of six different keying

Figure 28-3
The four units with
each link secured by
a different keying
variable

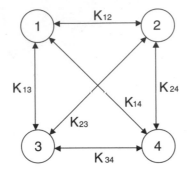

Figure 28-3
The four units with
each link secured by
a different keying
variable

variables. In addition, as the net grows to n members, it requires $\dfrac{n(n-1)}{2}$ different variables. For n large, this is on the order of n-squared.

Fortunately, we don't have to settle for one of these extremes. Other solutions are available. One solution is that of the *Key Distribution Center* (KDC). The KDC concept is sketched in Figure 28-4.

The KDC is a facility that is located in a highly secure hosted environment. As we will see, the security of the entire net depends on the KDC and its operations remaining inviolate.

Figure 28-4
The KDC and the
nodes it serves

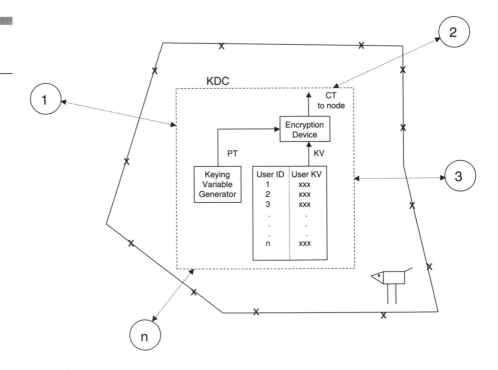

The KDC has the following components:

- A database, which is indexed by a user ID or node number and contains a secret and unique keying variable commonly held by only the user node and the KDC

- A keying variable generator, which is a device that generates a new and unique keying variable on demand

- An encryption device for secure communication with any user for whom a secret keying variable is on file in the KDC's database

Suppose nodes A and B want to converse securely, but no secret keying variable exists between them. One of them, let's say node A, calls the KDC and requests a secret keying variable for a secure session with node B.

The KDC generates a secret keying variable session key, K_{AB}, to use between nodes A and B. The KDC then looks up node A's secret keying variable in its database. It encrypts K_{AB} using this variable and sends the encrypted K_{AB} to node A. The KDC then looks up node B's secret keying variable in its database. It encrypts K_{AB} using this variable and sends the encrypted K_{AB} to node B. Nodes A and B can now conduct a secure session with both of them using K_{AB}, and the KDC is out of the loop.

Exercise 28

The purpose of this exercise is to examine the risk of network compromise as n grows and to help set the stage for the next module.

1. Assume we have n nodes and we are operating them with the same keying variable as we did with the four nodes of Figure 28-2. Also assume that the probability of a compromise at any particular node is p. Display the probability that the net will be compromised, $P_{compromised}$, versus n for n in the range $10 \leq n \leq 1,000$ for $p = 10^{-6}$ and $p = 10^{-4}$.

2. Assume you have a secret that you encrypt and publicly post the ciphertext. You use a 56-bit keying variable and then split the keying variable into two equal-size, nonoverlapping segments of 28 bits each. You give one of these segments to Trustee A and give the other to Trustee B. If one of these trustees tries to break the cipher (that is, recover the secret), how many keying variables would the trustee have to try on average in order to be successful?

Secret
Sharing

Secret sharing is a technique that can be extremely valuable operationally. When employed properly, it can provide an excellent operational feature and security system differentiator.

Problem 2 in Module 28, "Keying Variable Distribution," asks you to assess the wisdom of splitting a 56-bit cryptographic keying variable $(k_1, k_2, \ldots, k_{56})$ into two parts and giving each part to a different trustee. The motive behind this, presumably, is to require two persons to pool their respective knowledge in order to reconstitute the secret variable. When you solved the problem, you discovered that instead of enhancing system security, the practice actually diminished it as the entire keying variable could now be discovered by either trustee in 2^{27} trials (on average). Is there a better way to do this?

Suppose we had generated 56 random bits $(r_1, r_2, \ldots, r_{56})$ instead and given one trustee these bits and given the other trustee the 56-bit keying variable bit-by-bit mod 2 added to the random bits, that is, $(k_1 \oplus r_1, k_2 \oplus r_2, \ldots, k_{56} \oplus r_{56})$. Now both trustees would have a 56-bit quantity that does not enable them to shortcut trying 2^{55} trials on average to cryptanalyze a 56-bit key cipher. This is an example of a *secret-sharing system* or *shadow system*. In general, secret sharing enables a K-bit secret to be spread among T trustees so that any t of them, $t \leq T$, can recover the secret by pooling their knowledge, but any $t - 1$ of them cannot glean any information about the K-bit secret that would enable them to solve for it with less work than a nontrustee. For the previous system, $K = 56$, $T = 2$, and $t = 2$.

Now let's take a look at a slightly more involved version to get a sense of the architectural power that secret sharing can deliver to the designer. Suppose we have a 2-bit secret $(K = 2)$ and the secret bits are (k_1, k_2). Let us work with three trustees $(T = 3)$ and permit any two of the three trustees to reconstitute the secret bits $(t = 2)$. First, we must generate two random bits (r_1, r_2). Then,

- We give the two bits (r_1, r_2) to the first trustee.
- We give the two bits $(k_1 \oplus k_2 \oplus r_1, k_2 \oplus r_2)$ to the second trustee.
- We give the two bits $(k_2 \oplus r_1, k_1 \oplus r_2)$ to the third trustee.

Make sure you see that no trustee knows anything that would provide an advantage for solving for the secret bits.

Now consider the trustees two at a time. If the first and second trustees pool their knowledge, they can reconstitute (k_1, k_2). They can first add their second bits together, yielding k_2. They can then add k_2 to the sum of their first bits, yielding k_1.

If the first and third trustees pool their knowledge, they can also reconstitute (k_1, k_2). All they need to do is add their first bits together, yielding k_2, and then add their second bits together and recover k_1.

The second and third trustees can proceed by adding their first bits together, yielding k_1. They can then add k_1 to the sum of their second bits, yielding k_2.

You can begin to see how secret sharing can make all sorts of problems a lot easier. The combinations to a two-combination bank lock, for example, might be split among four people, two receiving one combination and two receiving the other. This way if one person is sick, there will be sufficient folks to open the vault. But what happens if two people are sick and they are two people who hold the same combination? We're out of luck. However, if we had used a secret-sharing system with $T = 4$ and $t = 2$, then any two of the four people could call in sick and the two people who showed up would be able to open the vault.

Another scenario concerns command and control. Consider that there are three levels of command—L1, L2, and L3—as shown in Figure 29-1. Assume that orders can be given by a higher level to a lower level.

The cryptoarchitect has set up a secret-sharing system for the command keys at each level with the parameters $T = 3$ and $t = 2$ so that, for any group at any level, so long as there are at least two trustees within a group, those trustees can reconstitute the key at the node to which they depend one level up.

Now suppose that a large strike decimates the forces. In fact, let's suppose it's a decapitation strike and the top level of command is gone, as shown in Figure 29-2.

Figure 29-1
The three levels
of command

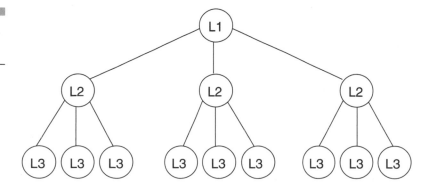

Figure 29-2
The remains after
the strike

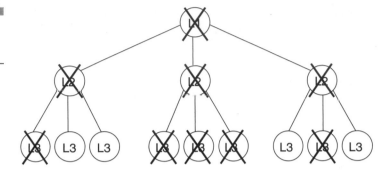

Figure 29-3
Reconstitution of two
L2 keys

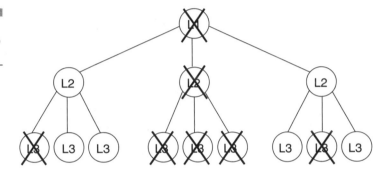

It looks pretty grim, but behold! The two remaining L3 units on the left reconstitute the L2 key at their superior node. A similar process occurs for the two remaining L3 units on the right. We end up with the result depicted in Figure 29-3.

The two reconstituted L2 levels can now reconstitute the L1-level key, and command control is reestablished for the remaining forces.

A general scheme for secret sharing is given by Kenneth Rosen.[1] The scheme can be implemented easily and enables construction for more general T and t values. It is applicable to integers, not just bits. The system is constructed as follows:

■ Define T and t.

■ Let the secret be denoted by the non-negative integer K.

---

[1]Kenneth Rosen, *Elementary Number Theory,* 4th edition, (Addison Wesley Longman, Inc., 2000).

- Select a prime p such that $p > K$.
- Select a set of T pairwise prime numbers $m_1, m_2, \ldots, m_T, (m_i, m_j) = 1$, $i \neq j$ such that

$$m_1 < m_2 < \cdots < m_T \tag{29.1}$$

and

$$m_1 m_2 \cdots m_t > p m_T m_{T-1} \cdots m_{T-t+2} \tag{29.2}$$

- Select a member, u, of the principal complete residue set mod(p) and create

$$K_0 = K + up \tag{29.3}$$

- Finally, create T shadows, $k_1, k_2, \ldots, k_T$, according to

$$k_1 \equiv K_0 \bmod(m_1)$$
$$k_2 \equiv K_0 \bmod(m_2) \tag{29.4}$$
$$\vdots$$
$$k_T \equiv K_0 \bmod(m_T)$$

Anyone in possession of t or more shadows can recover the secret, K. Anyone in possession of fewer than t shadows cannot. t or more shadows can be used to recover the secret, K, by solving for the smallest positive residue of K_0 modulo the product of the $\{m_i\}$ corresponding to the t shadows. This residue can be found using the Chinese Remainder Theorem and, after it is found, we can determine K by finding the largest integer x so that $K_0 - xp$ is non-negative and subtracting xp from K_0. An example will make this much clearer.

Let's assume that we want a system wherein $T = 3$ and $t = 2$ and that we are protecting a number, K, in the range $0 \leq K < 100$. Such an arrangement might correspond to one number of a safe combination distributed to three trustees so that any two might construct that number.

Following the previous system constraints, we pick

$$p = 101 \tag{29.5}$$

$$m_1 = 106 \tag{29.6}$$

$$m_2 = 109 \tag{29.7}$$

$$m_3 = 111 \tag{29.8}$$

$$u = 10$$

Now for the fun part. Let's assume that the secret number is

$$K = 42 \qquad \text{(29.9) [A galactic choice!]}$$

Therefore,

$$K_0 = 1052 \tag{29.10}$$

We now create the three shadows:

$$k_1 \equiv 1052 \bmod(106) = 98 \tag{29.11a}$$

$$k_2 \equiv 1052 \bmod(109) = 71 \tag{29.11b}$$

$$k_3 \equiv 1052 \bmod(111) = 53 \tag{29.11c}$$

Let's see if we can recover K from the first two shadows—k_1 and k_2. We have the simultaneous mixed radix equations:

$$K_0 \equiv 98 \bmod(106)$$
$$K_0 \equiv 71 \bmod(109) \tag{29.12}$$

Using the approach and notation of the Chinese Remainder Theorem, we compute inverses according to

$$\rho_1 M_1 \equiv 1 \bmod(106)$$

$$\rho_2 M_2 \equiv 1 \bmod(109) \tag{29.13}$$

The inverses are easily found to be

$$\rho_1 = 71$$

$$\rho_2 = 36 \tag{29.14}$$

From these values, we find that

$$K_0 \equiv (71 \times 98 \times 109 + 36 \times 71 \times 106) \bmod(11554)$$

$$\equiv 1052 \tag{29.15}$$

We then find the largest multiple of $p = 101$ that we can subtract from K_0 in Equation 29.15 without making the remainder negative. This multiple is 10 and we are left with

$$K = 1052 - 10 \times 101$$

$$= 42 \tag{29.16}$$

Exercise 29

The purpose of this exercise is to extend the $T = 3, t = 2$ case for a 2-bit secret variable K to $T = 4, t = 2$, and to further examine Rosen's secret-sharing technique.

1. We studied the $T = 3, t = 2$ case for a 2-bit secret variable K. Suppose we now have a fourth trustee to whom we give the 2 bits $(k_1 \oplus r_1, k_2 \oplus r_1 \oplus r_2)$. Show that we now have a 2-bit secret K in a secret-sharing system with $T = 4, t = 2$.

2. For the text's example of Rosen's secret-sharing technique, show that the secret K can be recovered from the other two pairs of shadows—k_1 and k_3, and k_2 and k_3.

Randomness II

In order to produce cryptovariables such as the secret keying variable or the initialization vector, it is often desirable to have a truly random source. You can choose from a number of candidates. The thermal noise associated with a resistor is an excellent candidate.[1] When using such sources, it is important to understand some of the theory behind the conversion of a random analog waveform into a series of bits, particularly the quantization threshold and the sampling rate to ensure sample-to-sample decorrelation. This module takes a look at some of these questions using simulated Gaussian noise.

The production of random bits to produce a keytext was one of the support systems required for SIGSALY, the voice encryption system that we looked at in Module 3, "Digitization of Plaintext." As J.V. Boone and R.R. Peterson report in the article on SIGSALY,[2]

> Key generation was a major problem. The basic requirements for the key (one essential part of the total encryption system) was that it should be completely random, and must not repeat, but could still be replicated at both the sending and receiving ends of the system.
>
> This was accomplished for SIGSALY by using the output of large (four-inch diameter, fourteen-inch high) mercury-vapor rectifier vacuum tubes to generate wideband thermal noise. This noise power was sampled every twenty milliseconds and the samples then quantized into six levels of equal probability. The level information was converted into channels of a frequency-shift-keyed (FSK) audio tone signal which could then be recorded on the hard vinyl phonograph records of the time.

To generate truly random, vice *pseudorandom* bits, we need to start with a random process such as thermal noise. In this module, we simulate such a spectrally wideband random noise source with pseudorandomly generated Gaussian noise. This is done by taking a series of uncorrelated samples of a Gaussian (or *normal*) random number generator. The sample values are created to have zero mean and unit variance.

In Figure 30-1, the top plot is a graph of 1,000 samples of the output of the Gaussian noise generator. The bottom plot is the result of converting the Gaussian noise samples, $\{g(n)\}$, to bits, $\{b(n)\}$, which are represented as -1s and $+1$s according to the following rule:

---

[1] W. Holman, J. Connelly, and A. Dowlatabadi, "An Integrated Analog/Digital Random Noise Source," *IEEE Transactions on Circuits and Systems-I: Fundamental Theory and Applications* 44, no. 6 (June 1997): 521–528.

[2] "The Start of the Digital Revolution: SIGSALY Secure Digital Voice Communications in World War II," op. cit.

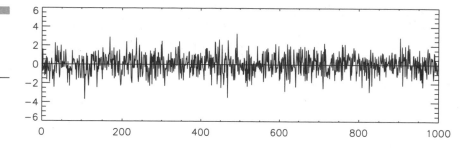

Figure 30-1
Uncorrelated
Gaussian samples
and derived bits

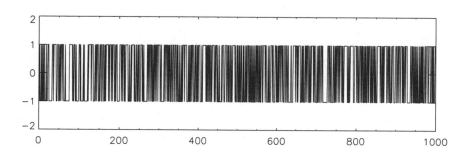

$$b(i) = \begin{cases} +1, & g(i) \geq \theta \\ -1, & g(i) < \theta \end{cases}$$
(30.1)[3]

where θ is a threshold. For the results in Figure 30-1, $\theta = 0$.

The number of -1 bits is 479 and the number of $+1$ bits is 521. Using the normal approximation to the binomial that we considered in Module 5, "What We Want from the Keytext," we see that the probability of this distribution of bits, 479/521, is within only 1.33 standard deviations of the mean of a binomial distribution for $n = 1,000$, $p = 1/2$. Therefore, we are inclined to accept that the bits have been produced by a balanced binary process with $p = 1/2$.

However, if we have not set our threshold properly, perhaps due to an electrical imbalance between the analog noise source and the hard quantizer, then we will have a different situation. Figure 30-2 is essentially the same as Figure 30-1 with one exception. The threshold for quantization in Figure 30-2 is not set to 0, but rather set to $\theta = 0.25$. In this case, the num-

---

[3]We trust that departing from our usual representation of bits as 0s and 1s to $+1$s and -1s will do no harm. It should make a couple of the computations a bit easier to handle.

Figure 30-2
Uncorrelated
Gaussian samples
and derived bits

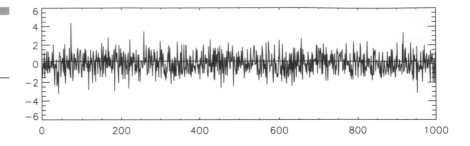

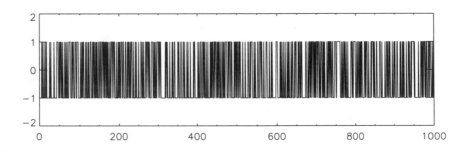

ber of -1 bits is 608 and the number of $+1$ bits is 392. This is a great difference. In fact, this particular outcome falls 6.83 standard deviations from the mean of a binomial distribution for $n = 1,000$, $p = 1/2$ and constitutes an event that inclines us to reject the hypothesis of balance.

An unbalanced binary source can still be used to approximate a balanced source. This can be done in a number of ways. An elegant solution is to use the results of the fifth problem of Exercise 4 in Module 4, "Toward a Cryptographic Paradigm," (von Neumann's coin). In this case, we take bit samples two at a time. We output a bit only if two members of the sampled pair of bits are different. We output a 0 if, say, the two samples are $-1, +1$ and output a 1 if the two samples are $+1, -1$.

Another approach is to smooth the unbalanced bit stream by mod-2 adding together successive samples of bits. Consider that we have a biased source of bits such that the probability of 1 is p and the probability of 0 is $1-p$. If we add n of the bits together mod-2, as shown in Figure 30-3, the sum will be a 1 with a probability of p_Σ, where

$$p_\Sigma = \frac{1}{2}\left(1 - (1 - 2p)^n\right) \qquad (30.2)$$

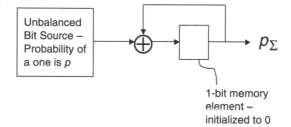

Figure 30-3
A sum for smoothing
an unbalanced bit
stream

Note that this approach does not eliminate a bias; in other words, there is no n such that p_Σ can be made to equal one-half if p is not one-half. However, this approach can be quite efficacious when mitigating even a severe bias. We plot p_Σ versus p for various n values in Figure 30-4.

We must consider another parameter. A correlation may exist between samples of the analog Gaussian process. A Gaussian noise source has a finite bandwidth, of course, and it may also be low-pass filtered. A model of such a source is sketched in Figure 30-5.

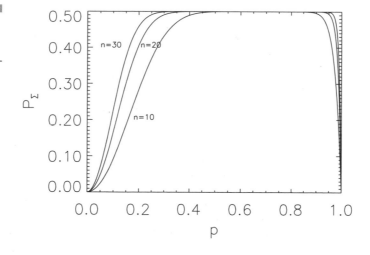

Figure 30-4
p_Σ versus p for $n =$
10, 20, and 30

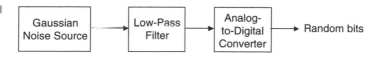

Figure 30-5
A mechanism for
producing random
bits

Figure 30-6
Correlated Gaussian
samples and
derived bits

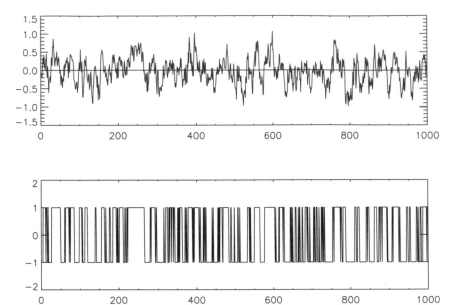

Figure 30-6
Correlated Gaussian
samples and
derived bits

In Figure 30-6, the top plot is a graph of 1,000 samples of the output of a low-pass-filtered Gaussian noise generator. The bottom plot is the result of converting the Gaussian noise samples, $\{g(n)\}$, to bits, $\{b(n)\}$, as we have done before. The wideband Gaussian noise, the noise before the low-pass filtering, is called *white noise* as its power spectral density is substantially flat over a wide frequency range. After filtering, the noise power spectral density has been shaped to something considerably different from substantially flat. The resulting noise is called *colored* as not all frequencies, or colors, are present equally.

White noise has the property that adjacent samples are uncorrelated. Indeed, if we calculated the expected value, $E(\cdot)$, of the adjacent Gaussian noise samples before low-pass filtering, we would expect $E(\cdot) = 0$. The low-pass filtering, however, insinuates an intersample dependence and, for the Gaussian samples in the top plot of Figure 30-6, we find that the *intersample correlation coefficient*, ρ_x, of the low-pass-filtered Gaussian samples is

$$\rho_x = \frac{E(x(n)x(n+1))}{E(x^2(n))} \tag{30.3}$$

$$= 0.82$$

As you might expect, the coloring of the noise has an effect on the bit stream. If we also calculate the *interbit correlation coefficient*, ρ_b, we find

$$\rho_b = E(b(n)b(n+1))$$

$$= 0.582 \tag{30.4}$$

The effect of the interbit correlation can be dramatically seen if we count the 500 di-bits in the bottom plot of Figure 30-6. Table 30-1 shows the results of the count.

You can see that the effect of coloring the Gaussian noise by low-pass filtering has made the di-bit counts depart from a flat distribution over the four possible cases to one that favors the two cases where the bits forming the di-bit pair are the same. The bits have become highly correlated.

There is, incidentally, a straightforward relationship between the inter-sample correlation coefficient, ρ_x, of the low-pass-filtered Gaussian samples and the interbit correlation coefficient, ρ_b. The relationship is known as the *arcsine* law[4]:

$$\rho_b = \frac{2}{\pi} \sin^{-1}(\rho_x) \tag{30.5}$$

This is plotted in Figure 30-7.

Table 30-1	**Di-bit**	$(-1, -1)$	$(-1, +1)$	$(+1, -1)$	$(+1, +1)$
Frequency distribution of the 500 di-bits of Figure 30-6	**Di-bit frequency**	212	50	52	186

[4]For instance, see A. Papoulis, *Probability, Random Variables, and Stochastic Processes*, 2nd edition, (New York: McGraw-Hill, 1984).

Figure 30-7
The arcsine law

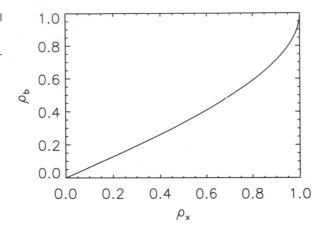

Exercise 30

The purpose of this exercise is to gain a deeper understanding of the smoothing technique for biased bits.

1. Derive Equation 30.2. Hint: Consider the expansions of

$$((1 - p) - p)^n \tag{30.6}$$

and

$$((1 - p) + p)^n \tag{30.7}$$

using the binomial theorem.

2. Consider the shape (see Figure 30-4) of the p_Σ versus p curves if n is even. What would the shape of the p_Σ versus p curve be if n is odd?

Cryptovariables

The cryptovariables make the state of a cryptographic system unique. They are at the heart of the system's security and as such must be understood and properly managed from their creation through their deployment into the system.

Cryptovariables comprise those entities that put the publicly known cryptographic algorithm into a secret state so that it can serve to protect the traffic entrusted to it. The cryptovariables we have already studied include

- The secret keying variable

- The *initialization vector* (IV)

Both of these variables are extremely important for ensuring the security of the cryptosystem. Both variables possess unique qualities and require different and specific handling. For example, the secret keying variable must be protected from public view. It is the heart of the security of the cryptosystem. If it is compromised, it must be assumed that any traffic encrypted using that cryptovariable is compromised because, as we have mentioned previously, the details of the cryptosystem are presumed to be publicly known.

The IV, on the other hand, is generally not considered secret. Indeed, it is often transmitted as part of the preamble to an enciphered message. However, if it is misused, the IV can lead to message compromise. This could happen, for example, by using the same IV for two different messages encrypted using the *output feedback* (OFB) mode, creating a depth situation. For this reason, the IV must be created in such a way that the reuse of an individual IV is extremely unlikely. In many cases, the IV does not have to be unpredictable; it just cannot be repeated.

The keying variable, on the other hand, often serves many messages. It does not need to be changed for every message like the IV. However, the keying variable should be essentially unpredictable in all cases. If we generate our keying variables by using some of the techniques discussed in Module 30, "Randomness II," we must be sure that we are approximating a balanced binary Bernoulli source. Let's look at the consequence of not doing this.

Suppose we had a machine that required only a $k = 4$-bit secret keying variable. Let's suppose that we generated these variables with a binary Bernoulli but unbalanced source so that the probability of 1 was $p < 1/2$. How would a cryptanalyst attack such a system assuming that no shortcut to the solution was available and that the only attack, therefore, was to posit each possible cryptovariable until the solution appeared?

The cryptanalyst would try each possible cryptovariable, but not in just any order. Some cryptovariables are more likely than others. For example,

Figure 31-1
Order of
cryptovariable
exhaustion

CRYPTOVARIABLE	
0000	Zero Ones
0001	
0010	One One
0100	
1000	
0011	
0101	
0110	Two Ones
1001	
1010	
1100	
0111	
1011	Three Ones
1101	
1110	
1111	Four Ones

Figure 31-1
Order of cryptovariable exhaustion

the *a priori* probability of the cryptovariable 0000 is $(1 - p)^4$, which is greater than the probability of the cryptovariable 1111 whose *a priori* probability is p^4.

Figure 31-1 presents a logical order in which to test all the possible cryptovariables so the solution can be obtained more quickly than through a randomly ordered exhaustion.

Figure 31-1 has the cryptovariables partitioned into five classes according to their number of ones, or *weight*, as it is sometimes called. For $p < 1/2$, we try all cryptovariables in the order of increasing weight.[1]

We can look at the effect of $p < 1/2$ with respect to real-world values of k by calculating the expected number of trials for the solution using the preceding intelligent approach. In Figure 31-2, we calculate the effect of using $k = 56$-bit keying variables constructed from a binary Bernoulli source for which $0.25 \leq p \leq 0.5$. The calculations were done so that the effect was

---

[1]It might not be obvious how to efficiently generate all $\binom{k}{w}$ k-bit keying variables of weight w for large values of k. Fortunately, there is algorithmic help. See J. Bitner, G. Ehrlich, and E. Reingold, "Efficient Generation of the Binary, Reflected Gray Code and Its Applications," *Communications of the ACM* 19 (1976): 517–21.

Figure 31-2
Equivalent number of
keying variable bits

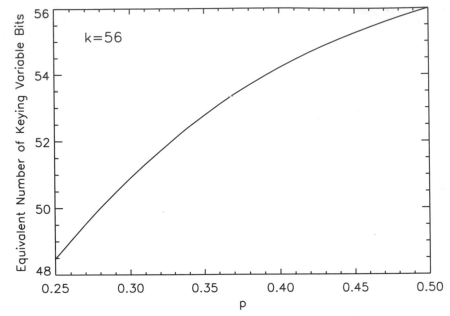

translated into the equivalent number of keying variable bits produced with $p = 0.5$.

The effect of producing keying variable bits with $p \neq 0.5$ may not seem dramatic, but the number of equivalent bits drops to about 55 when $p = 0.44$. This is a 50 percent reduction in exhaustive cryptanalysis. Therefore, if we were expecting to gain a solution in one year if the cryptovariables were produced from a balanced binary Bernoulli source, we could expect to wait only half of a year if $p = 0.44$. That is quite a reduction.

Exercise 31

The purpose of this exercise is to stimulate thought. Weaknesses in a cryptosystem do not always need to be in the cryptoalgorithm. In fact, misuses of the cryptosystem, such as the use of an inappropriate protocol, can obviate the need for a sophisticated, computer-intensive attack.

1. You run the ENCRYPTION4U service company. Your company provides a secret transmission service from point A to point B. This is done by using a block cipher function in the OFB mode. The communications uses a simple link-layer protocol that sends a packet that looks like

IV	MESSAGE

   The customer stops by at point A and his or her recipients show up at point B. The customer presents a message at point A, which is encrypted and transmitted by your company to point B over a communications link that may be monitored. You believe that you are adequately protecting the keying variable used and the messages are properly delivered. The IV for OFB does not need to be kept secret; indeed, it is often sent in the clear as it is in your service. Furthermore, in order to give your customers the sense of having more control over their messages, you offer to let them specify the IV that will be used for their messages.

   A customer comes to you and reports that he suspects his messages are being compromised by your service. How could this be happening? Perform a comprehensive audit of all the possibilities and weigh their respective likelihoods.

2. Security depends on, or rather is a result of, many different activities working properly together. It is not enough to generate, distribute, account for, and properly destroy keying variables; they must also be properly entered into the cryptographic device that they are serving. If the variable is entered incorrectly, it may cause the device to be keyed with an incorrect keying variable and one that may be cryptographically weak.

   The *Data Encryption Standard* (DES) was hardened somewhat against this contingency. The 64-bit keying variable consists of only 56 bits that specify internal cryptographic settings within a keyed DES. The other

8 bits each form parity checks over the eight 7-bit blocks of the 56 key-ing variable bits. Thus, a DES device can be built so that it detects any single error in any of the eight 7-bit blocks. However, the construction of this goes another step. The mandated parity checks have odd parity bits. This requires an odd number of ones in the total number of ones in each 7-bit block in addition to its associated parity bit. For example, if the first 8 bits of the keying variable are designated by

$$kb_1, kb_2, kb_3, kb_4, kb_5, kb_6, kb_7, kb_8 \qquad (31.1)$$

with kb_8 as the odd parity bit, kb_8 is computed as

$$kb_8 = kb_1 \oplus kb_2 \oplus kb_3 \oplus kb_4 \oplus kb_5 \oplus kb_6 \oplus kb_7 \oplus 1 \qquad (31.2)$$

Why is the parity bit scheme constructed like this? What does it do for the system security?

3. What fraction of the total possible 2^k k-bit keying variables (k is even) is contained in the subset of keying variables that have exactly $k/2$ zeros and $k/2$ ones? Use Stirling's approximation.

4. A cryptosystem consists of a *central processing unit* (CPU) and a coprocessor in a secure enclosure, as shown in Figure 31-3. It contains a port where ciphertext can enter and exit in addition to a physically secure port where plaintext can enter and exit. It also has one data channel that is allowed to exit the enclosure. The purpose of this data port is to enable remotely based technicians to monitor the coprocessor use. The technicians can tell when the coprocessor is active.

Figure 31-3
A monitored
cryptosystem

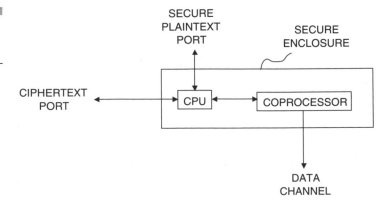

The coprocessor is used for exponentiation in a Diffie-Hellman public key cryptosystem. The coprocessor calculates

$$\alpha^x \bmod(y) \tag{31.3}$$

What are the security implications of this system?

Crypto-obsolescence

The fifth part of this book comprises only two modules. It addresses a difficult and pressing question—what, if anything, can we do about the older cryptosystem, and when can we do it? Consider that a cryptoprinciple is developed and it is immune to the most powerful cryptanalysis of the day. It is impressed into service. At that point, it begins to age, but the cryptanalytic arts do not halt. They become stronger, more refined, and increasingly dangerous to the aging cryptoprinciple. What can be done? An upgrade or replacement seems like a logical answer, but if the cryptoprinciple has been popular and if it is widely used, particularly in a hardware incarnation, then the question of immense disruption is raised as different networks or parts of networks change their cryptosystems and become incompatible even if it is just for a short while.

We look at a few of these issues, which are called the axes of attack. I also present a somewhat personal solution that I unabashedly proffer for consideration.

The Aging Cryptoalgorithm

As algorithms age, they offer less protection. This is because they remain fixed and are surrounded by the inexorable progress of the cryptanalytic arts and new capabilities. In this module, we take a look at this problem in general terms.

A cryptographic algorithm has three enemies, or three-dimensional axes of attack. They are illustrated in Figure 32-1 and include

- Newly discovered cryptanalytic algorithms
- Ever more powerful computers including computational hardware and memory
- New and better architectures for performing computational tasks

Each of these three dimensions can prove to be a threat, but problems most likely come following advances along a plurality of these axes.

Let's look at an example to help explain this. Suppose that we are trying to cryptanalyze a simple substitution cipher. Our faster-than-exhaustion cryptanalytic algorithm is to make a frequency count of each of the S different symbols used in the plaintext-ciphertext alphabets. Assume the cryptogram has a total of n symbols. For each of the n symbols, we must

- Decode the symbol.
- Increment the counter counting the frequency of occurrence of the particular decoded symbol.

Let's measure the combination of both of the previous two tasks with one fixed time unit: τ_{inc}. Doing this on a single processor then requires T_1 time units where

$$T_1 = n\tau_{inc} \tag{32.1}$$

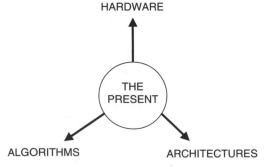

Figure 32-1
The three axes of attack

Figure 32-2
The first
multiprocessor
architecture

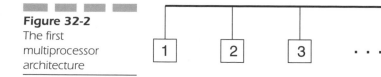

Now suppose that we try to speed up our task by applying more than one processor, say, P processors. How will we build such a multiprocessor farm and what algorithm will we use?

For our first choice, we pick a linear array of the P processors that are all connected via a backbone bus, as shown in Figure 32-2.

Each processor is assigned only n/P of the symbols and each processor maintains its own frequency of occurrence table. At the conclusion of the counts, these tables must be combined. As only one processor can transmit over the commonly shared bus at one time, the algorithm for combining the P individual frequency of occurrence tables is to send the table from processor 1 to processor 2, combine the two tables, send the result from processor 2 to processor 3, combine the two tables, and so on until the grand total is available at processor P. If it takes τ_{xmit} time units to transmit a table over the bus and τ_{add} time units to add two tables together, then the time it takes the P processors to have the grand frequency of occurrence table available at processor P, T_P, is

$$T_p = \frac{n}{P}\tau_{inc} + (P - 1)(\tau_{xmit} + \tau_{add}) \tag{32.2}$$

The *speedup* afforded by using P processors instead of just one processor is denoted by S_P and is defined as

$$S_p = \frac{T_1}{T_P} \tag{32.3}$$

For the architecture, algorithm, and hardware of Figure 32-2,

$$S_P = \frac{n\tau_{inc}}{\frac{n}{P}\tau_{inc} + (P - 1)(\tau_{xmit} + \tau_{add})} \tag{32.4}$$

$$= \frac{P}{1 + \frac{P(P - 1)}{n}\left(\frac{\tau_{xmit} + \tau_{add}}{\tau_{inc}}\right)}$$

If

$$r = \frac{\tau_{xmit} + \tau_{add}}{\tau_{inc}} \tag{32.5}$$

we have

$$S_P \cong \frac{P}{1 + \dfrac{P^2}{n} r} \tag{32.6}$$

There is a simple and insightful model for multiprocessor speedup.[1] This model seeks to caste the speedup, S_P, in the following form:

$$S_P = \frac{P}{(1 - \alpha)P + \alpha + P\sigma(P)} \tag{32.7}$$

where

- α is the fraction of the computational problem that may be parallelized or solved concurrently on the P processors.
- $\sigma(P)$ represents the overhead required to manage a parallel solution. This term includes synchronization and interprocessor communication functions.

If we map Equation 32.6 into Equation 32.7, we see that according to this model, our problem may be completely parallelized as $\alpha = 1$, but significant overhead appears as

$$\sigma(P) = \frac{P}{n} r \tag{32.8}$$

If we could reduce $\sigma(P)$, we could increase the speedup we get from using P processors. If we look at the architecture of Figure 32-2 and the algorithm we are using, we see that the chained passing and adding of the intermediate frequency of occurrence tables in order to form the grand frequency of occurrence table is straightforward in design, but not parallel in nature. Can we do better?

---

[1] B. Buzbee, "The Efficiency of Parallel Processing," *Los Alamos Science* (Fall 1983): 71.

Figure 32-3
The second multiprocessor architecture

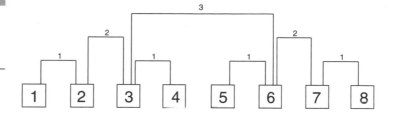

If we could add some of the intermediate frequency of occurrence tables simultaneously, it would augur a reduction in $\sigma(P)$. However, in order to modify our algorithm to do this, we must also modify the computational architecture since the processors currently share a common bus and no more than one processor can transmit on that bus at any one time. So we modify our architecture to that of Figure 32-3. (In Figure 32-3, we show an example where P is a power of two. Our algorithm is best suited for this, but it will work for other cases as well.)

This second multiprocessor architecture functions as follows. The processors connected by busses bearing the label 1 combine their frequency of occurrence tables simultaneously. To do this, processor 1 sends its table to processor 2 where it is combined and awaits further action. Simultaneously, processor 4 sends its table to processor 3 for combining, processor 5 sends its table to processor 6 for combining, and processor 8 sends its table to processor 7 for combining.

After the label 1 bus activity has ceased and the tables have been combined, the table contents are passed on bus level 2. Processor 2 sends its table to processor 3 for combining, and processor 7 sends its table to processor 6 for combining.

The final step takes place after label 2 bus activity has ceased and the tables have been combined. At that point, processor 6 sends its table to processor 3 over bus level 3. The grand table is then produced and resides in processor 3.

The time it takes this new architecture and algorithm to produce the grand frequency of occurrence table is

$$T_P = \frac{n}{P}\tau_{inc} + (\log_2 P)(\tau_{xmit} + \tau_{add}) \qquad (32.9)[2]$$

---

[2]Again, please remember that in our example P is a power of 2.

This new architecture and algorithm gives us the greater speedup of

$$S_P = \frac{P}{1 + \dfrac{P\log_2 P}{n} r} \tag{32.10}$$

where now

$$\sigma(P) = \frac{\log_2 P}{n} r \tag{32.11}$$

Of course, depending upon hardware conditions and advances, the factor r may change and affect the speedup.

A factor that may be considered by itself is memory. The state-of-the-art in memory is important and its advances spell significant mileposts for certain classes of attack. One of these attacks is the *meet-in-the-middle attack*. Consider that we have a double encryption using block cipher functions, as illustrated in Figure 32-4.

Assume that we have some matched plaintext-ciphertext $\{PT, CT\}$ pairs and that the two keying variables (K_1 and K_2) are each of length k bits.

Figure 32-4
Double encryption

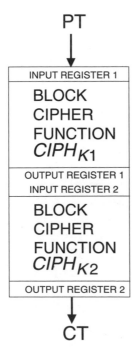

We could try to solve for the two keys by trying all possible keying variables. This would require around 2^{2k} tests.

However, we could also try a very different approach. We could try all possible 2^k values of K_1 and all 2^k values of K_2 in the following manner. We encrypt a known plaintext under each value of K_1 and store the content of output register 1 associated with each particular value of K_1. We also decrypt the ciphertext corresponding to the known plaintext under each value of K_2 and store the content of input register 2 associated with each particular value of K_2. Next we look for matches between the stored values of output register 1 and the stored values of input register 2. A match identifies a candidate (K_1, K_2) solution pair.

The candidate solution pair may be checked by seeing whether other matched plaintext-ciphertext pairs are consistent with the candidate (K_1, K_2) solution pair.

This attack trades off solution time for memory. Instead of solving an order $O(2^{2k})$ computational problem, we are faced with solving only an order $O(2^k)$ computational problem, but we require memory on the order of $O(2^k)$.

An analogy might make this a bit clearer. Suppose that we had to solve the maze problem of Figure 32-5. The goal is to get from Vertex A to Vertex B. The rules for traveling in the maze are simple. You can move along a line as long as you don't cross an X. Also, you may cross the shaded bar in the middle only once.

Well, how do we solve this? We could try all possible paths and indeed we would find a solution, as illustrated in Figure 32-6. But how might we use a meet-in-the-middle approach?

Figure 32-5
A maze problem

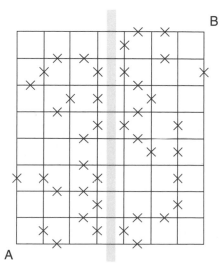

Figure 32-6
A solution to the
maze problem

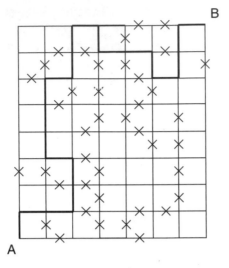

Figure 32-7
The maze pulled
apart

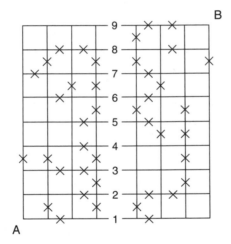

A meet-in-the-middle approach can be done very simply. First we pull the maze apart and number the cross links, as shown in Figure 32-7.

Next we start from Vertex A and find and record all of the cross links that can be reached from Vertex A. The reachable cross links are contained in the set $\aleph_1$ where

$$\aleph_1 = \{1, 6, 7, 8, 9\} \qquad (32.12)$$

We do the same thing from Vertex B and find that from Vertex B, we can reach the cross links in the set $\aleph_2$ where

$$\aleph_2 = \{2, 3, 4, 5, 8\} \qquad (32.13)$$

The intersection of $\aleph_1$ and $\aleph_2$ is the set {8} so we know that a solution is available for getting through the maze. The solution must travel across cross link 8, which we see is the case in Figure 32-6.

But the story of computation doesn't end here. In fact, it's probably just beginning. The future will write itself, but if you'll permit me to write just a few more paragraphs, we'll take a peek at something truly extraordinary.

The security of much of the public key cryptographic world seems to rest on factoring being a very difficult problem for increasingly large composite numbers.

In 1903, Frank Nelson Cole delivered a remarkable address to the *American Mathematical Society* (AMS). In the words of Anthony W. Knapp,[3]

> He was a master lecturer. It is said that his 1903 address "On the factoring of large numbers" to the Society was met with a standing ovation after he lectured without saying a single word, multiplying two large integers[4] and verifying that their product was $2^{67} - 1$, a number that Mersenne had thought should have been prime.

Since then, the following has occurred:

- **Reported in 1975**   Morrison and Brillhart factored $2^{128} + 1$ into two primes—one of 17 digits and the other of 22 digits.

- **Reported in 1981**   Brent and Pollard factored $2^{256} + 1$ into two primes—one of 16 digits and the other of 62 digits.

- **Reported in 1984**   Simmons, Davis, and Holdridge factored $2^{251} - 1$ into three primes—one of 21 digits, one of 23 digits, and one of 26 digits.

- **Reported in 1989**   A 100-digit number was factored.

- **Reported in 1994**   A 129-digit number was factored.

---

[3]A. Knapp, "Frank Nelson Cole," *Notices of the AMS* 46, no. 8 (1999): 860.
[4]193707721 and 761838257287

And so it goes. More powerful algorithms are found, hardware becomes faster, and multiprocessing is engaged. However, we may be standing on the threshold of yet another revolution—*quantum computing*. Quantum computation is still a gleam in the eye of physicists, engineers, and applied mathematicians, but it appears as though it may come to pass. If it does, it will be tremendous because, by the nature of quantum mechanics, it will be possible to parallelize a problem's solution through the superposition of quantum states. In other words, the upshot of the previous hardly illuminating statement is that a quantum computer's states can keep up with the exponential growth of some very difficult mathematical problems such as factoring. It will be exciting to watch. The future isn't what it used to be.

Exercise 32

Multiprocessors have been suggested for years for a number of cryptographic attacks. According to the published papers, they seem to be conceptually useful for handling those problems that are completely partitionable. This short exercise asks you to look at a model for such behavior.

1. Multiprocessors for cryptanalysis have been talked about for years. One of the earliest, if not the earliest, articles was written by Diffie and Hellman.[5] They postulated a machine for solving for a *Data Encryption Standard* (DES) keying variable. Their machine, which was late 1970s vintage, would have required about a million chips, cost about $20 million, and have exhausted the entire 2^{56} keying variable space in about half a day.

   More recently, Gershon Kedem and Yuriko Ishihara[6] reported that they had used a *Single Instruction Multiple Data* (SIMD) machine to solve for UNIX passwords. An SIMD is an array of multiprocessors that is an excellent architecture for problems that can be *parallelized*, that is, divided so that the individual pieces all run in approximately the same time and do not have much, if any, interdependence.

   We have already looked at the model in Equation 32.7 for the multiprocessor speedup of a problem's solution. If no interprocessor communication exists, that is, if $\sigma(P) = 0$, then we have the simpler model, which is known as Ware's model,[7] where

   $$S_P = \frac{P}{(1 - \alpha)P + \alpha} \qquad (32.14)$$

   The speedup is critically dependent upon the fraction of the problem, α, that can be parallelized. Graph S_P as a function of α for $P = 10$ processors.

---

[5]W. Diffie and M. Hellman, "Exhaustive Cryptanalysis of the NBS Data Encryption Standard," *Computer* 10, no. 6 (1977): 74–84.

[6]Gershon Kedem and Yuriko Ishihara, "Brute Force Attack on UNIX Passwords with SIMD Computer," Proceedings of the 8th USENIX Security Symposium, August 23, 1999, Washington D.C.

[7]B. Buzbee, "The Efficiency of Parallel Processing," op cit.

SUPERDES™

As cryptoalgorithms reach the end of their useful life due to cryptanalytic advances, the difficult question is raised of what to use next and how and when to introduce the new system considering the time, effort, and expense of replacing the old. The Data Encryption Standard (DES) has been a remarkably successful algorithm and has been used for decades. Replacing it will not be easy or painless. This module looks at a modest proposal from a pending patent that was developed by Noel D. Matchett and me. This proposal describes a family of algorithms named SUPERDES.[1] We believe that the proposal is interesting and worth studying as it highlights the problems of an actual aging cryptoalgorithm and suggests a way to move forward with less pain to the existing infrastructure.

We have touched on the DES a number of times. To refresh your memory, it is a cipher block function that was promulgated in 1977 by the National Bureau of Standards via the Department of Commerce's vehicle: *Federal Information Processing Standards* (FIPS) No. 46.

In its most basic form, the *electronic codebook* (ECB) mode, the DES works with 64-bit *input/output* (I/O) blocks, $b = 64$. Encryption is under the control of a $k = 56$-bit secret keying variable, K. The DES is perhaps the most widely analyzed cryptographic system in history and has stood well against many different cryptanalytic attacks. It has served as a model for the development of many other cryptographic algorithms and has been widely employed, which has created a problem.

Much of its employment has been in hardware, particularly in hand-held mobile privacy radios. This represents a very large investment and presence in the DES engine. Replacing all of the units would be extremely costly and highly disruptive because replacement would almost certainly have to be performed in stages, which often creates interoperability problems.

By nature, many communication theoreticians tend to be less sensitive to the operating rhythm of communications and more zealous toward instituting the best of the state-of-the-art features. It is suggested that this is the case with the DES and its replacement. In this module, we look at a way to move forward to greater security, which is a necessity among serious cryptographers, but use a strategy that provides backwards algorithmic compatibility so the transition can be made smoother for those charged with ensuring operational effectiveness.

DES security has declined due to the inexorable advance in the available worldwide computer power coupled with the fame of the DES algorithm.

---

[1]SUPERDES is a trademark of and is used with the permission of Information Security Incorporated who negotiates SUPERDES licenses on behalf of the patent application owners.

Challenges have been mounted through parallel exhaustive attacks and so-called special attacks in which an individual seeks to find a path to a solution that is computationally less than that of simple exhaustion.

Two seminal publications examine the cryptanalysis of the DES algorithm. These publications represent two powerful and distinct cryptanalytic approaches. Neither approach was initially successful in defeating the DES, but both approaches deserve consideration as genres of potent cryptanalysis. The first of these was reported in a paper by Diffie and Hellman.[2] This paper discussed the construction of a large parallel processor in which the entire 56-bit keying variable space was partitioned over a very large number of identical independent processors. The paper also advanced the argument that declining computation costs would eventually reduce the cost of a solution to a nominal sum.

This type of attack can, of course, be countered by increasing the size of the keying variable, and a keying variable not much greater than 56 bits could effectively frustrate this approach.

Eli Biham and Adi Shamir detail the second attack in a lengthy paper.[3] This paper introduces a statistical cryptanalytic method termed *differential cryptanalysis*, which the authors describe as "a method which analyzes the effect of particular differences in plaintext pairs on the differences of the resultant ciphertext pairs." These differences can be used to assign probabilities to the possible secret keying variables and suggest the most probable variables.

Biham and Shamir used the DES as an example for their cryptanalytic method. They characterized the DES as an iterated cryptosystem (a series of rounds) in that it realizes a strong cryptographic function by iterating a weaker function (a single round) many times. Their attack is based on Boolean differencing for which the structure of the DES appears to be an ideal candidate.

When applied to the DES, their attack would have beaten the exhaustion of the possible keying variables if the DES had used less than 16 rounds of iteration. In particular, Biham and Shamir made the following observations:

1. The modification of the key scheduling algorithm cannot make the DES much stronger. (The key schedules are the 16 48-bit vectors

---

[2]W. Diffie and M. Hellman, "Exhaustive Cryptanalysis of the NBS Data Encryption Standard," *Computer* (June 12, 1977): 74–84.

[3]Eli Biham and Adi Shamir, "Differential Cryptanalysis of a DES-like Cryptosystem," The Weizmann Institute of Science, Department of Applied Mathematics, June 18, 1990.

formed from the 56-bit secret keying variable. Their use is depicted in Figures 10-4 and 10-5 in Module 10, "The Block Cipher Function—A Modern Keystone to the Paradigm.")

2. The efficacy of the attacks on the DES with 9 to 16 rounds is not influenced by the P permutation. Likewise, the replacement of the P permutation by any other fixed and publicly known permutation or function cannot make them less successful. (The P permutation is a permutation of 32 bits and is also depicted in Figure 10-5.)

3. The replacement of the order of the S-boxes without changing their values can make the DES weaker. (See Figure 10-5 and the accompanying discussion in the first problem of Module 10.)

4. The replacement of the exclusive-or operation by the more complex addition operation makes the DES much weaker.

5. It is easy to break the DES if randomly chosen S-boxes are used. Even a change of one entry in one S-box can make the DES easier to break.

Additional work related to differential cryptanalysis encompasses so-called linear cryptanalysis[4] and other statistical attacks. Biham and Shamir[5] published an improvement of one of these other statistical attacks and reported breaking the full 16-round DES faster than through an exhaustive search. The statistical attack required a large volume of matched plaintext and ciphertext.

The various cryptographic attacks and the increase in computer power available to exhaust the 56-bit keying variable have caused the U.S. government to recommend *triple DES*. Triple DES essentially uses the 16 rounds of the single DES engine three times with different keying variables to provide increased security. The penalty paid for using triple DES is a threefold increase in running time over the single DES. As the government notes,[6]

> With regard to the use of single DES, exhaustion of the DES (that is, breaking a DES encrypted ciphertext by trying all possible keys) has become increasingly more feasible with technology advances. Following a recent hardware-based DES key exhaustion attack, NIST can no longer support the use of single DES for many applications. Therefore, government agencies with

---

[4]M. Matsui, "Linear Cryptanalysis Method for DES Cipher," *Abstracts of EUROCRYPT'93*, W112–23.

[5]Biham and Shamir, "An Improvement of Davies' Attack on DES," *EUROCRYPT'94*, 461–7.

[6]FIPS PUB 46-3, October 25, 1999.

legacy systems are encouraged to transition to triple DES. Agencies are advised to implement triple DES when building new systems.

Noel D. Matchett and I[7] proposed a way to upgrade cryptographic security by creating an improvement to the DES algorithm so that it could be continually strengthened yet retain a backwards compatible mode in order to serve legacy requirements. We believed that because the basic DES chassis had been analyzed for so many years and continued to prove its design soundness, it made sense to build upon it rather than scrap it for a new algorithm whose cryptanalytic history is too young to provide high confidence that conversion is worth the investment. The goal was to find some way for DES to be strengthened, but still have the strengthened version run in the same time as a single DES.

An additional goal was to develop a backward compatible mode, enabling the very high investment in DES hardware to be retained and interoperable with the newer and more secure algorithm. The only disadvantage we see in triple DES today is the threefold penalty in speed over DES. However, we consider IBM's concept of triple DES and its capability of backward compatibility with DES to be one of the outstanding cryptographic concepts in the last 25 years.

It is worth noting that the backward compatibility of cryptosystems is not a new consideration. According to A. Ray Miller,[8]

> Later in the war, however, the German Navy adopted a variant version of the Enigma cipher machine which used four rotors . . . the fourth rotor did not move; once the Enigma operator had set the initial rotation position by hand, it remained constant for the duration of the message. . . . This had a nice side effect, however. In practice, the fourth rotor and its new reflector had wiring chosen such that in one particular orientation the combination had exactly the same effective wiring as reflectors built for three rotor Enigmas. This gave the four rotor machines the ability to still communicate with the older three rotor machines.

One of the algorithm's features is the modification of the fixed permutation P block of the classic DES algorithm. (Again, see Figure 10-5.) The P permutation is applied after the S-boxes and the SUPERDES version

---

[7] Each inventor *a priori* blames the other for missing any cryptanalytic attacks that might be discovered that significantly reduce the security afforded by SUPERDES.

[8] A. Miller, "The Cryptographic Mathematics of Enigma," Center for Cryptologic History, National Security Agency, Fort George G. Meade, 11, 12, and 15.

preserves the permutation nature of the block; that is, it remains a one-to-one mapping. Replacing the known P permutation with a different known permutation will not augment the security of the modified DES as recounted in Point 2 of Biham and Shamir's results mentioned previously. The difference is that the SUPERDES proposal uses a new secret permutation, $P*$, for the P block. This has a default option to the original P permutation for backwards compatibility. The proposal also suggests that the new $P*$ permutation may be dynamically changed from round to round without incurring a time penalty. The proposal suggests that the new permutation module be constructed of a logical array of binary switches in a structured class of networks in order to construct permutations that can be changed quickly. Depending on the particular network implemented, a related fixed permutation can be computed so that when the binary switches are all set to a default condition, the resulting permutation created by the network when followed by the related fixed permutation results in a permutation that is equivalent to the original P permutation to create the feature of backwards compatibility.

The variable permutations can be based upon elements of the keying variable or can depend on additional elements such as an encipherment counter, frame counter, or some permanently fixed bits. The permutation selection can also be done by supplying the SUPERDES algorithm with an externally provided pseudorandom bit source, as sketched in Figure 33-1.

Figure 33-1
SUPERDES algorithm with an externally provided pseudorandom bit source

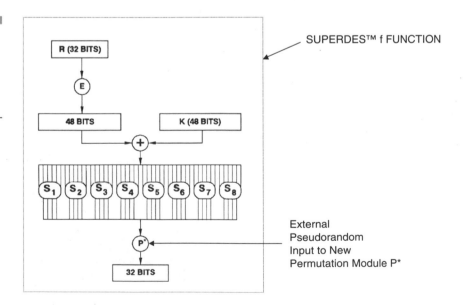

SUPERDES™ f FUNCTION

R (32 BITS)

E

48 BITS   K (48 BITS)

$+$

S_1 S_2 S_3 S_4 S_5 S_6 S_7 S_8

$P*$

32 BITS

External Pseudorandom Input to New Permutation Module $P*$

Figure 33-2
The β-element in its
two states

This external source could be from another cryptographic engine. This feature enables the SUPERDES algorithm to be continually upgradable without suffering a decrease in operating speed.

We can build powerful permutation networks out of a spatial array of interconnected two-state building blocks termed β-elements. A β-element is sketched in Figure 33-2 in both of its states.

The β-element has two states, and the state is determined by the control bit, C. When $C = 0$, the two input ports I1 and I2 are connected to the two output ports O1 and O2, respectively. When $C = 1$, I1 and I2 are connected to O2 and O1, respectively.

Consider that we have $N = 2^n$ input ports and the same number of output ports. Therefore, $N!$ distinct permutations exist between I1 ... IN and O1 ... ON. The P permutation of the DES is one of these permutations for $n = 5$.

The *omega network* can effect a significant subset of the $N!$ permutations. An omega network consists of n columns or stages of 2^{n-1} β-elements each, wired together in a special way. This special way is a simple pattern, which is termed a *perfect shuffle*,[9] that is replicated between columns. The pattern will become clear after we look at the example provided in Figure 33-3.

The omega network has a simple and regular architecture. Building this type of network for $N = 32$ to serve as part of the $P*$ permutation is straightforward.

But how many permutations and what sorts of permutations can be performed by such a network? Figure 33-3 shows that we have a total of 12 β-elements in the omega network. Each β-element is set by a single bit so no more than $2^{12} = 4,096$ permutations are possible. For eight distinct elements, there are, of course, $8! = 40,320$ possible permutations so we see that an omega network as we have defined it can accomplish only a small

---

[9]H. Stone, "Parallel Processing with the Perfect Shuffle," *IEEE Transactions on Computers* 20 (1971): 153–61.

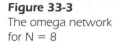

Figure 33-3
The omega network
for N = 8

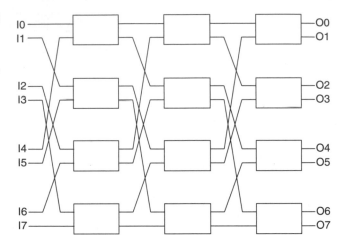

fraction of these. In general, an omega network for which $N = 2^n$ can do the following:

- Map any single Ij to any Ok.
- Perform the identity permutation.
- Perform the permutation where input Ix is mapped to output $O(ax+b)\mathrm{mod}(2^n)$, where a is odd.
- Be upwardly subsumed; in other words, any permutation doable by an omega with $N = 2^n$ is embedded in the permutations doable by an omega network with $N = 2^{n+1}$.

So, if we were to use an omega network with $N = 32$ as part of the P^* network, we could simply follow it with the P permutation. That way when all of the omega network control bits were set to zero, the P^* permutation would become the P permutation and we would have backwards compatibility.

Other permutation networks are available and some of them can realize all $N!$ permutations. For example, an extended omega network using $3n - 1$ stages instead of just n stages can be used to perform all permutations between 2^n inputs and 2^n outputs.[10]

---

[10]C-L. Wu and T.-Y. Feng, "The Universality of the Shuffle-Exchange Network," *IEEE Transactions on Computers* 30 (1981): 324–32.

One very efficient permutation network that can implement all the possible permutations is known as the *Beneš–Waksman network*.[11] This network has a structure that is not regular in the sense of the omega network, but it is still generally quite easy to construct and reduce to a high-speed electronic implementation.

Figure 33-4 shows the Beneš–Waksman network for eight inputs. Note that it requires 17 β-elements and therefore 17 bits to set it up. In the usual employment of an omega network, a Beneš–Waksman network, or any other *space division connecting network*, the network is used to permute input ports to output ports in order to connect parties, not just form a permutation. In the SUPERDES case, however, it is not necessarily a particular permutation that is being sought; rather, it could be a selectable or dynamic permutation in order to greatly magnify the complexity of the cryptanalysis needed to successfully attack the improved algorithm.

The Beneš–Waksman network is considered efficient because as $N \to \infty$, it uses the fewest possible number of β-elements to build a connecting network that is capable of implementing any of the $N!$ possible permutations. Let the number of β-elements required for a Beneš-Waksman network connecting $N = 2^n$ I/O ports be denoted by $S(n)$. It is known that

$$S(n) = 2^n(n - 1) + 1 \qquad (33.1)$$

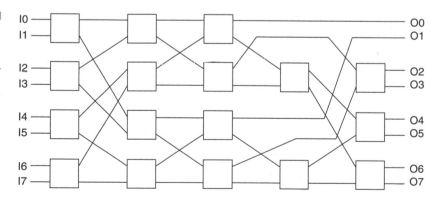

Figure 33-4
The Beneš–Waksman network for $N = 8$

[11]V. Beneš, *Mathematical Theory of Connecting Networks*, (Academic Press, 1965) and A. Waksman, "A Permutation Network," *Journal of the Association for Computing Machinery* 15 (1968): 159–63.

Now let $B(n)$ be the number of bits required to specify one of $(2^n)!$ permutations. Therefore,

$$B(n) = log_2((2^n)!) \tag{33.2}$$

We see that

$$\lim_{n \to \infty} \frac{S(n)}{B(n)} = 1 \tag{33.3}$$

Exercise 33

This exercise highlights an interesting aspect of an omega network.

1. Consider an omega network of $N = 2^n$ inputs and outputs. Let us assume that a packet of data enters at input port i, where $0 \leq i \leq 2^n - 1$, and we desire that it be routed to output port j, where $0 \leq j \leq 2^n - 1$.[12] Let us equip the packet with an n-bit header with bits j_{n-1}, $j_{n-2}, \ldots, j_0$ where these bits are the normal binary representation of j. In other words,

$$j = j_{n-1}2^{n-1} + j_{n-2}2^{n-2} + \cdots + j_0 \tag{33.4}$$

Therefore, the packet is equipped with the binary address of the output port on which it wants to exit. When entering input port i, the bits of the binary representation of i, that is, $i_{n-1}, i_{n-2}, \ldots, i_0$ where

$$i = i_{n-1}2^{n-1} + i_{n-2}2^{n-2} + \cdots + i_0 \tag{33.5}$$

are bit-by-bit mod-2 added to the bits of the binary representation of j. This sequence of bits is denoted as $c_{n-1}, c_{n-2}, \ldots, c_0$ and is

$$c_{n-1} = i_{n-1} \oplus j_{n-1} \tag{33.6}$$

$$c_{n-2} = i_{n-2} \oplus j_{n-2}$$

$$\vdots$$

$$c_0 = i_0 \oplus j_0$$

The sequence of bits $c_{n-1}, c_{n-2}, \ldots, c_0$ are the control bits of the β-elements through which the packet will pass to the desired output port. They should be used in the order from the most significant bit, c_{n-1}, to the least significant bit, c_0.

Try this technique with a packet entering on port 3 and directed to output port 5 in an omega network with $N = 8$.

---

[12] Please pay close attention to the labeling convention for the input ports illustrated in Figure 33-3.

PART VI

Channel-Based Cryptography

The sixth and final part of this book also comprises only two modules. The phrase *channel-based cryptography* is one that I coined myself. I believe that a class of cryptography exists in which the channel itself is, or is a major part of, the cryptovariable. You may have heard people speak of the *subliminal channel*. This was a term coined by Gus Simmons. It refers to a channel that carries additional information in a fashion other than by its main function. For example, many of you are familiar with a safe's combination lock. For the combination of, say, 10-20-30, you turn the dial to the left a number of times until you come to 10, stop, and reverse the direction of the dial in preparation for entering 20. Did you know that many safes open under their combination regardless of whether you start to the left or right? It's an interesting feature and has potentially great value as a subliminal channel if a violent criminal forces you to open the safe, for example. You comply, but you have two choices. You can start from the left or right and, if you have previously affixed a silent alarm to be triggered on one direction but not the other, you can signal your duress and still comply with the felon's command.

We look at two channels in this module: a suburban radio channel that can be made to manufacture a symmetric cryptovariable for two parties and the quantum cryptographic channel. The latter channel is still largely in the research realm, but it shows promising signs of vitality.

The Channel Is the Cryptovariable

This module is an example of channel-based cryptography that may be of use in a suburban area that contains a lot of traffic and buildings. The privacy of the system depends on this complex scenario. The system enables two users to develop the same secret quantity between each other so it is technically a symmetric type of cryptography, but it does not require a priori secret agreement. Also, much of the signal and mathematical processing hardware required may already be part of a communications receiver.

The realization of reliable cellular communications was one of the great accomplishments of the past few decades. The problem was quite challenging in suburban areas because of the prevalent and severe multipath. Reflections caused a signal to be received with time spread that was often to the point of severe intersymbol interference. Reflections off of specular and diffuse surfaces that are in motion in relation to the transmitter or receiver also induce Doppler spreads that further complicate the received signal.

The one bright spot is that the channel is reciprocal, albeit for a very short time. A signal sent from A to B is altered in the same manner as a signal sent from B to A so long as the two measurements are made almost simultaneously. Wait too long, or move just a short distance, and the channel changes significantly.

This type of channel has a reciprocal, measurable, and quantifiable randomness that enables us to create a cryptosystem.[1,2] We are concerned with the urban *Ultra High Frequency* (UHF) channel and our example will be for frequencies near 1,900 MHz. The system functions as follows:

1. Party A transmits a set of T tones to B. These tones have publicly known phases and amplitudes between each other. Party B measures[3] the magnitudes and phases of the received tones. Party B could easily do this by taking a Fourier transform.

2. Party B measures the phases between each of the tones, quantizes the phase differences to L different levels, and notes the magnitudes of the individual received tones.

3. Party B then immediately transmits the same set of tones, with the same relative phases and amplitudes that Party A transmitted.

---

[1] J. Hershey, A. Hassan, and R. Yarlagadda, "Unconventional Cryptographic Keying Variable Management," *IEEE Transactions on Communications* 43, no. 1 (1995): 3–6.

[2] A. Hassan, W. Stark, J. Hershey, and S. Chennakeshu, "Cryptographic Key Agreement for Mobile Radio," *Digital Signal Processing* 6 (1996): 207–12.

[3] The parties may wait a short interval, perhaps around 10 microseconds, in order for the significant multipath to stabilize, that is, for all significant rays to arrive.

4. Party A then does what Party B did; it measures the phases between each of the tones, quantizes the phase differences to L different levels, and notes the magnitudes of the individual received tones.

5. Both parties then have approximately the same string of (T-1) L-ary symbols. This string of symbols is then put into a minimum distance decoder and the information bits of the code word estimated by the decoder become the secret quantity held by both parties. This secret quantity may then be used either as keytext or, more likely, as a keying variable for another cryptosystem.

Let's consider a simple example. Suppose we transmit a suite of eight tones with intertone spacing so that the adjacent tone phases become uncorrelated. We form the seven differences between the adjacent tones starting from the tone of the lowest *radio frequency* (RF) and hard-quantize the phase differences so that an absolute phase difference that is less than or equal to $\pi/2$ radians is represented by a 0 and an absolute phase difference greater than $\pi/2$ radians is represented by a 1. We then decode the 7 bits just derived by using the Hamming (7,4) code to produce a secret 4-bit keying variable. (For this particular code, we will not use the magnitudes of the received tones in the decoding process.)

Assume that we discover the data shown in Table 34-1, where $\theta_{i+1} - \theta_i$ is the difference in radians between the adjacent tone phases and $Q(\cdot)$ is the quantizer function. The quantized phase differences, reading down the right-most column of Table 34-1, constitute the tone suite binary vector. This vector is 0011011.

Table 34-1

Example data

i	$\theta_{i+1} - \theta_i$	$Q(|\theta_{i+1} - \theta_i|)$
1	-0.2	0
2	1.3	0
3	3.1	1
4	-2.9	1
5	1.5	0
6	-1.9	1
7	2.8	1

Table 34-2

The Hamming (7,4) code words and their matches to the tone suite binary vector

Keying Variable Bits				Parity Bits			M
k_1	k_2	k_3	k_4	p_1	p_2	p_3	
0	0	0	0	0	0	0	3
0	0	0	1	1	1	1	5
0	0	1	0	1	1	0	4
0	0	1	1	0	0	1	6
0	1	0	0	1	0	1	2
0	1	0	1	0	1	0	4
0	1	1	0	0	1	1	4
0	1	1	1	1	0	0	3
1	0	0	0	0	1	1	4
1	0	0	1	1	0	0	2
1	0	1	0	1	0	1	3
1	0	1	1	0	1	0	5
1	1	0	0	1	1	0	1
1	1	0	1	0	0	1	3
1	1	1	0	0	0	0	2
1	1	1	1	1	1	1	4

In Table 34-2, we show the 16 code words of the Hamming (7,4) code. This particular code is known as a *perfect code* because it completely packs the entire code space so that each code word is at a minimum *Hamming distance* of 3 bits from any other code word; in other words, each code word differs from every other code word by at least 3 bits. We also tabulate the number of bit matches, M, between each of the 16 code words of the Hamming (7,4) code to the 7-bit tone suite binary vector.

We can decode this by finding the largest match M. It is 6 and we declare the secret keying variable to be k_1, . . . , k_4 of the code word for which M is maximum. Therefore, the secret keying variable is 0011.

The previous example is for illustration only. Decoding the Hamming (7,4) code is not normally done in this manner. We just wanted to clearly illustrate the concept of searching for the code word with a minimum distance to the tone suite binary vector.

We can put this channel-based cryptosystem on more solid ground by appealing to a model conceived by William Jakes.[4] This model calculates the quantity λ^2 where

$$\lambda^2 = \frac{J_0^2(\omega_m \tau)}{1 + s^2 \sigma^2} \tag{34.1}$$

where ω_m is the maximum Doppler shift between transmitter and receiver, τ is the transmission delay, s is the RF frequency $\omega_2 - \omega_1$, and σ is the time delay spread resulting from multipath. Jakes' model suggests that the intertone phases are decorrelated when λ^2 is less than or equal to 0.4.

Guidance for estimating the multipath delay, σ, can be derived from a number of sources. Michael J. Gans[5] suggests that the multipath delay is around one-fourth of a microsecond in some suburban areas and 5 microseconds in some urban areas. A report from NTIA's Institute for Telecommunication Sciences[6] concurs and adds a curious wrinkle to the suburban case:

> Mobile impulse response measurements were taken in the 1,850–1,990 MHz band in three different macrocellular (cell radii of 5 km) environments: flat rural, hilly rural, and urban high-rise. Spatial diversity with a 15-wavelength separation was employed by using a dual-channel receiver. All antennas were omnidirectional and vertically polarized. The data were analyzed to provide delay statistics; spatial diversity statistics; multipath power statistics; number of paths, path arrival time, and path power statistics; and correlation bandwidth statistics. The urban high-rise cell showed more multipath components (out to 4 or 5 μs in delay) than the rural cells. Very long delays (greater than 10 μs), while not seen often, were seen more frequently in the rural cells than in the urban high-rise cell.

---

[4]William Jakes, Jr., *Microwave Mobile Communications*, (Wiley, 1974), Section 1.5.4.

[5]Michael J. Gans, "A Power-Spectral Theory of Propagation in the Mobile-Radio Environment," *IEEE Transactions on Vehicle Technology* VT-21 (1972): 27–38.

[6]J. Wepman, J. Hoffman, and L. Loew, "Impulse Response Measurements in the 1850–1990 MHz Band in Large Outdoor Cells," NTIA Report (1994): 94–309.

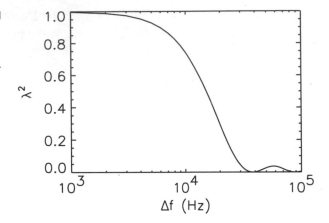

As an example, in Figure 34-1, we plot λ^2 versus the RF separation between two tones for two tones, one at 1,900 MHz and the other at 1,900 MHz $+ \Delta f$ with $\sigma = 5$ microseconds and a Doppler shift resulting from about 5 miles per hour. As you can see, Δf does not need to be very large in order to cause phase decorrelation.

███ ███ Exercise 34

This exercise asks you to calculate a solution to a simple multipath problem.

1. Assume that the two signals

$$\sin(\omega t) \tag{34.2}$$

$$A \sin(\omega t + \theta) \tag{34.3}$$

arrive at an omnidirectional antenna. The antenna sees the signal

$$y = \sin(\omega t) + A \sin(\omega t + \theta) \tag{34.4}$$

which is indistinguishable from the signal

$$y = B \sin(\omega t + \phi) \tag{34.5}$$

Find B and ϕ.

The Quantum Cryptographic Channel

Quantum cryptography (QC) is a fascinating contemporary research area. Unlike algorithms of public key cryptography, whose strength is known to be no greater than certain classical problems (for example, factoring large, hard composite numbers), QC depends on what most scientists have come to believe as a law of nature. In this module, we look at the background of QC, considering elementary quantum theory, and then discuss the application.

QC is a system that can be used for keying variable distribution on an optical channel such as provided by free space and some fiber-optic infrastructures. The security depends on a basic principle of quantum mechanics. A working proof-of-principle demonstration was reported in 1989.[1]

As we will see, QC depends on several technologies for its efficient and secure operation. The first of these is the capability to produce a single photon of a specified polarization or phase. The production of less than a single photon (none) adversely affects implementation efficiency. The production of more than a single photon seriously affects the security of the process.

Photons come from light sources and most light sources exhibit Poisson statistics. According to this statistical model, a Poisson light source produces μ photons per basic time interval, for example, within a second or laser pulse, and the probability that the source will produce exactly y photons in the basic time interval, $P_Y(y)$, is

$$P_Y(y) = e^{-\mu} \frac{\mu^y}{y!} \tag{35.1}$$

It has been suggested that single photons might be produced by highly attenuating a laser pulse.[1] By properly choosing and ensuring the multiple parameters of such an arrangement, you can achieve the production of single photons about 10 percent of the time, no photons about 90 percent of the time, and, most importantly, since it affects the system's security, multiple photons less than 1 percent of the time.

In order to better understand QC, it is necessary to have a working model of a single photon. A long line of progress in physics has suggested that single photons and other atomic particles have wavelike properties so it is natural to express their observable behavior as though they were

---

[1]C. Bennett, F. Bessette, G. Brassard, L. Salvail, and J. Smolin, "Experimental Quantum Cryptography," *Journal of Cryptology* 5 (1992): 3–28.

describable by a wave function,[2] $\Psi(x, t)$.[3] In order to be consistent with the observable behavior, the following form was indicated for $\Psi(x, t)$:

$$\Psi(x, t) = Ae^{-2\pi/h\,(Et - p_x x)} \tag{35.2}$$

where

- A is the amplitude of the wave function.
- E is the energy of the particle.
- h is Planck's constant.
- p_x is the momentum of the particle in the x direction.

Note that we can form

$$\frac{\partial \Psi}{\partial x} = \frac{2\pi i}{h} p_x \Psi \tag{35.3}$$

Therefore, we can write a momentum operator as

$$p_x = \frac{h}{2\pi i} \frac{\partial}{\partial x} \tag{35.4}$$

and compute the expected value of the momentum as

$$\bar{p}_x = \int \Psi^* \frac{h}{2\pi i} \frac{\partial}{\partial x} \Psi dx \tag{35.5}$$

(The asterisk indicates the complex conjugate.) The expected position is found by a similar application of the position operator, x, as

$$\bar{x} = \int \Psi^* x \Psi dx \tag{35.6}$$

An operator, α, representing a physical quantity such as the momentum or position is called a *dynamical variable*, which is a term introduced by Dirac. The expectation of such a variable described by the wave function Ψ is

$$\bar{a} = \int \Psi^* a \Psi dv \tag{35.7}$$

where the integral is taken over the volume appropriate to Ψ.

---

[2] Edwin Goldin, *Waves and Photons*, (New York: John Wiley and Sons, 1982).

[3] We'll work in one spatial dimension only.

If we define $\alpha = p_x x$, then

$$\frac{h}{2\pi i} \frac{\partial}{\partial x} (x \Psi) = \frac{h}{2\pi i} x \frac{\partial \Psi}{\partial x} + \frac{h}{2\pi i} \Psi \qquad (35.8)$$

But if we set $\alpha' = x p_x$, then we have

$$x \frac{h}{2\pi i} \frac{\partial}{\partial x} (\Psi) = x \frac{h}{2\pi i} \frac{\partial \Psi}{\partial x} \qquad (35.9)$$

We note that

$$p_x x - x p_x = \frac{h}{2\pi i} \qquad (35.10)$$

This means that the operators p_x and x do not commute.

An operator associated with a dynamical variable is termed *observable*. If an operator is observable, it implies that the operator is *Hermitian*—that is, its own adjoint. Only Hermitian operators produce real eigenvalues.[4] We just looked at the product of the two dynamical variables, x and p_x, and noted that the difference in Equation 35.10 is nonzero. Therefore, $x p_x$ is not a dynamical variable, but if we had considered $x p_y$ or xy, then we would have commutativity.

The next step is to identify $\Psi^* \Psi$ as a probability density function. David Bohm[5] provides an excellent motivation by first requiring that a proper probability density function, $p(x)$, for, say, the particle position between x and $x + dx$ is given by $p(x)dx$ and that the basic and intuitive criteria required of such a probability density function is naturally satisfied by

$$p(x) = \Psi^*(x)\Psi(x) \qquad (35.11)$$

as

- $p(x)$ must be non-negative.

- The magnitude of the probability density function must, in some fashion, track the magnitude of $\Psi(x)$.

---

[4] Goldin, op. cit.

[5] David Bohm, *Quantum Theory*, (Englewood Cliffs, NJ: Prentice-Hall, 1951).

■ $p(x)$ must be invariant to independent variables, that is, variables whose effect on the probability density function may be dismissed on physical grounds.

■ $\int_{-\infty}^{+\infty} p(x)dx$ must be unity.

Bohm proceeds from his tentative identification in Equation 35.11 and shows that for the cases of interest to us, it is the correct choice.

The boundary conditions imposed on the wave function result in a set of *eigenvalues* and associated *eigenfunctions*. To determine the probability of any discrete eigenvalue, for example, α_n, we write Ψ as a linear combination of its eigenfunctions

$$\Psi = \sum_i c_i \Psi_i \tag{35.12}$$

and evaluate

$$\alpha_n = |c_n|^2 \int \Psi_n^* \Psi_n dv \tag{35.13}$$

The beauty and wisdom of QC is in the consideration of a single photon. This is where the wave theory of light is abandoned and the quantum world is entered. The key is found in the question of the assemblage or mixture of so-called pure states.

The Schroedinger equations, which describe the evolution of steady-state behavior of the wave function, are linear and thus admit to a superposition of solutions.

If we were to (mentally) prepare a mixture of pure cases (in other words, a mixture of different wave functions), what could or couldn't we say about the defining wave function of the mixture assemblage?

Following Edwin Kemble's development,[6] we see that the individual pure case wave functions are $\Psi_1, \Psi_2, \ldots$, with mixture weights $w_1, w_2, \ldots$, where

$\sum_i w_i = 1$, and can write

$$\bar{\alpha} = \sum_i w_i \int \Psi_i^* \alpha \Psi_i dv \tag{35.14}$$

---

[6]Edwin Kemble, *The Fundamental Principles of Quantum Mechanics*, (New York: McGraw-Hill, 1937).

where $w_i = |c_i|^2$. We can now ask an obvious but critical question: Can a mixture of cases, an assemblage, be considered equivalent to a single case?

Kemble proceeds by assuming that we can do this and discovers a contradiction as the corresponding wave function for the assemblage is then

$$\Psi = \sum_i c_i \Psi_i \qquad (35.15)$$

However, Ψ must also properly represent the moments of the mixture. Therefore,

$$|\Psi|^2 = \sum_i w_i |\Psi_i|^2 \qquad (35.16)$$

but this is generally inconsistent with the implication of Equation 35.15 that

$$|\Psi|^2 = \left| \sum_i w_i^{1/2} \Psi_i \right|^2 \qquad (35.17)$$

Therefore, a mixture cannot generally be considered as a single or pure case.

I often say, somewhat tongue in cheek, that a particle is a wave packet seeking a boundary condition. Eugene Wigner wrote about the orthodox view of measurement as follows:[7]

> The possible states of a system can be characterized, according to quantum-mechanical theory, by state vectors. These state vectors—and this is an almost verbatim quotation of von Neumann—change in two ways. As a result of the passage of time, they change continuously according to Schroedinger's time-dependent equation—this equation will be called the equation of motion of quantum mechanics. The state vector also changes discontinuously, according to probability laws, if a measurement is carried out on the system. This second type of change is often called the reduction of the waveform.

However, in QC, as in most practical applications of quantum theory, an observation must eventually be made. Thus, we are ineluctably drawn into a measurement process. We encounter the reduction of the wave function,

---

[7]E. Wigner, "The Problem of Measurement," *American Journal of Physics* 31 (1963): 6–15.

or what has also been called *wave packet collapse*, in the measurement process because one of the eigenvalues appears during this process—in other words, the probability of that particular eigenvalue becomes unity in the a posteriori. As Kemble puts it, " . . . regarding the measurement of a dynamical variable α . . . A successful, exact, individual measurement of α, whether retrospective or predictive, must always yield as its numerical result one of the eigenvalues of α."

The sea change between a multiplicity of possible eigenvalues and a single eigenvalue brought about by a measurement has prompted a serious and prolonged debate. It is not necessary, fortunately, for us to enter the debate even if we are willing and able. It is, however, important for us to realize that so long as a measurement is not made, the Schroedinger equation can be thought of as describing a deterministic evolution of Ψ and, as such, all sorts of helpful things such as superposition obtain. As Roger Penrose claims, "Whenever we 'make a measurement,' . . . , we change the rules . . . ."[8] It is strictly the measurement "that introduces uncertainties and probabilities into quantum theory." The importance of this will be clear when we look at the various optical structures and techniques for implementing QC.

The theory is interesting, but without a practical implementation, it is not helpful for engineering a system. As we will soon understand, the production of a single photon is a key implementation requirement for achieving efficiency and security. However, the production of a single photon has traditionally been very difficult to do on demand. Again, one technique that can be used for QC is to generate light pulses that have a small probability that they consist of more than one photon. This policy will, however, result in most pulses carrying no photons at all.

As mentioned previously, it is usual to model a pulsed light source as emitting photons according to Poisson statistics 35.1 for the number of photons in a single pulse. Using 35.1, we can write the probability that there will be two or more photons in a laser pulse, $P_Y(2^+)$, as

$$P_Y(2^+) = 1 - P_Y(0) - P_Y(1) \qquad (35.18)$$

$$= 1 - e^{-\mu} - \mu e^{-\mu}$$

$$\approx \frac{\mu^2}{2} \, for \, \mu \ll 1$$

---

[8] Roger Penrose, *The Emperor's New Mind*, (New York: Oxford University Press, 1989).

For small μ, the probability that a pulse comprises no photons is much more likely, by a factor of μ^{-1}, than the probability that a single photon comprises the pulse. If we constrain $P_Y(2^+)$, then

$$P_Y(1) \backsim \sqrt{2P_Y(2^+)} \qquad (35.19)$$

Table 35-1 enumerates $P_y(1)$ under this constraint for a few values of $P_Y(2^+)$.

One method for achieving a Poisson source with $\mu \ll 1$ is to severely attenuate the output of a source with significantly greater μ such as what is depicted in Figure 35-1.

An attenuator that can be modeled as transmitting an incoming photon with the probability p (and therefore absorbing a photon with the probability $1 - p$) where p is constant and memoryless (where the absorption

Table 35-1

$P_Y(1)$ constrained by $P_Y(2^+)$

$P_Y(2^+)$	$P_Y(1)$
10^{-2}	0.141
10^{-3}	0.0447
10^{-4}	0.0141
10^{-5}	0.00447
10^{-6}	0.00141

Figure 35-1
New photon source with $\mu \ll 1$

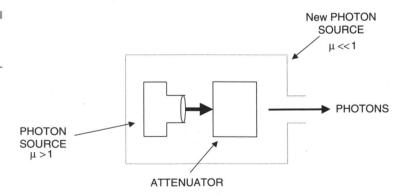

process is Bernoulli) has a fortuitous behavior when used with a Poisson source.

Following Gabor Herman's observations,[9] we can define $p_n(m)$ as the binomial probability that exactly m photons out of n are transmitted, so

$$p_n(m) = \frac{n!}{m!(n-m)!}p^m(1-p)^{n-m} \tag{35.20}$$

The probability, $p_X(x)$, that exactly x photons are transmitted from the new photon source of Figure 35-1 is

$$P_X(x) = \sum_{y=x}^{\infty} P_Y(y)p_Y(x) \tag{35.21}$$

$$= \sum_{y=x}^{\infty} e^{-\mu} \frac{\mu^y}{y!} \frac{y!}{x!(y-x)!} p^x(1-p)^{y-x}$$

$$= e^{-\mu} \frac{p^x}{x!} \sum_{y=x}^{\infty} \frac{1}{(y-x)!} \mu^y(1-p)^{y-x}$$

$$= \frac{e^{-\mu p}(\mu p)^x}{x!}$$

So, the statistic describing the new photon source of Figure 35-1 is also Poisson with the expected value μp.

Recently, great progress has been made in developing single photon emitter sources. One approach has been through using a technology called *quantum dots*. An article on quantum dots reports,[10]

Recent progress in lithography, colloidal chemistry, and epitaxial growth have made it possible to fabricate structures in which carriers or excitons are confined in all three dimensions to a nanometer-sized region of a semiconductor. Structures like these are commonly called quantum dots . . .

With self-assembled quantum dots, quantum control of carrier injection and photon generation is now possible. . . . Especially notable is the key role that quantum dots are likely to play in the emerging field of quantum-information

---

[9] Gabor Herman, *Image Restoration from Projections*, (Academic Press, 1980).

[10] P. Petroff, A. Lorke, and A. Imamoglu, "Epitaxially Self-Assembled Quantum Dots," *Physics Today—Online* 54, issue 5 (2001): 46.

science—either as building blocks where quantum information is stored in the spin degrees of freedom, or as a source of single photons for quantum communication.

Okay, we've looked at a bit of quantum theory and believe we can generate single photons. How do we make a cryptographically interesting system out of all this?

The key to answering this question lies in the measurement process. If we pick two nondynamical variables such as x and p_x, we will have non-commutativity as illustrated in Equation 35.10. This noncommutativity is, of course, at the heart of Heisenberg's famous principle. The construction of a QC system proceeds by exploiting this. As taught in Bennett's paper,[11]

> The essential quantum property involved, a manifestation of Heisenberg's uncertainty principle, is the existence of pairs of properties that are incompatible in the sense that measuring one property necessarily randomizes the value of the other. For example, measuring a single photon's linear polarization randomizes its circular polarization, and vice versa. More generally any pair of polarization states will be referred to as a basis if they correspond to a reliably measurable property of a single photon, and two bases will be said to be conjugate if quantum mechanics decrees that measuring one property completely randomizes the other.

The scheme starts to emerge. Using the two conjugate bases of linear polarization and circular polarization, we can send 1 bit of information with one photon if we have prior knowledge of which basis is used for the photon's generation. Let's say that linear polarization is being used. In this case, we will use a detector that channels the incoming photon into one of two optical paths—one path for vertically polarized photons and the other for horizontally polarized photons. Each path terminates in a detector. By noting the detector output, we will be able to decode 1 bit of data. For example, a horizontal polarization is a 0 and a vertical polarization is a 1.

If circular polarization is used, we will similarly use a detector that channels the incoming photon into a right-hand circular polarization optical path or a left-hand circular optical path. Again, each path ends in a detector and 1 bit of information is detected.

But suppose we don't know if the incoming photon is vertically polarized or circularly polarized. If we use a circular polarization handedness detector on a photon that is linearly polarized, then we will detect a right-hand

---

[11]C. Bennett et al, "Experimental Quantum Cryptography," op cit.

circular polarization with a probability of one-half and a left-hand circular polarization with a probability of one-half. No correlation will exist between the detected handedness of the circular polarization detectors and the state of the linear polarization. The situation is the same should we attempt the detection of circular polarization handedness with a pair of linear polarization detectors. This uncertainty result is equivalent to a succinct quantum mechanical argument set forth by William Wootters and Wojciech Zurek.[12]

A QC system then depends on a transmitter that successively transmits n photons, switching randomly between one of two conjugate bases and a detector that detects incoming photons with each photon detected by a detector selected randomly from one of the two conjugate bases. After transmitting the n photons, the receiver informs the transmitter of the basis used for each photon and the transmitter then informs the receiver regarding the correctness of those choices. From this knowledge, the receiver can determine what detections have information and what detections have no information.

A QC architect must specify many parameters of the system such as

- The size of n
- The minimum number of correct decisions required of the receiver
- The use of the receiver's correct decisions to encode information beyond individual 0s and 1s
- The protocol for handling an active attack, such as the man-in-the-middle attack, wherein an interloper makes a random detection, just as the intended receiver would, and sends a photon to the receiver that is randomly generated in one of the conjugate bases

Interest in QC is spreading and the supporting technology is growing. The U.S. Patent 5,966,224 "Secure Communications with Low-Orbit Spacecraft Using Quantum Cryptography" was issued on October 12, 1999. Under the field of the invention, we find

> As earth satellites and other spacecraft become even more prolific, issues relating to the security of earth to satellite communications have become exceedingly important. Particularly, satellite telemetry is becoming more susceptible to eavesdropping even as the importance of satellite intelligence and the security of data and of command paths to spacecraft is continually increasing. Current methods for securing satellite telemetry transmissions

---

[12] W. Wootters and W. Zurek, "A Single Quantum Cannot Be Cloned," *Nature* 299 (October 28, 1982): 802–3.

against third party interception rely on the perceived difficulty of interception of periodically uploaded random number generation "seeds" for use in crypto-logics in the spacecraft and at a secure ground station. However, these methods are open to eavesdropping, and are not demonstrably secure . . .

The present invention provides a system for secure telemetry between a low-orbit satellite and a secure ground station utilizing quantum cryptography based on nonorthogonal polarization states of pulsed laser light, attenuated to a level of one photon per pulse. These polarized photons are used to generate shared key material between the satellite and the ground station through optical systems which facilitate long-range communication through space.

It is therefore an object of the present invention to provide secure communication between a spacecraft in low-earth orbits and earth.

It is another object of the present invention to utilize quantum cryptography techniques to provide secure communication between a spacecraft in low-earth orbits and earth.

It is yet another object of the present invention to provide secure communication between a spacecraft in low-earth orbit and earth using polarized photons of light to transfer random key material.

▬ ▬ Exercise 35

QC may eventually become a valuable cryptographic method. It still faces a lot of development. This short exercise asks you to consider two aspects of QC implementation.

1. What issues would arise concerning the use of QC in long-haul optical fiber systems?

2. QC has been proposed for use in free-space channels. A notable example of this is the patent just cited. In free space, there is the question of light noise; in other words, there are a lot of photon emitters. In order for a QC system to work in a free-space channel, it must be highly probable that the detectors can successfully ignore or gate out unwanted photons. This can be done to some extent with spectral filtering and very tight time gates. This latter technique is widely advocated. In many cases, it is suggested that a very bright, ultrashort pulse precede the single photon transmission. This short, bright pulse of light may be used for synchronization. This is the case with U.S. Patent 5,966,224 in which we find in the summary of the invention:

> To achieve the foregoing and other objects, apparatus for generating a key to be used for secure communication between a spacecraft in a low-earth-orbit and an earth station comprises a laser for outputting pulses of light, with means for splitting the pulses of light into preceding bright pulses and delayed attenuated photon pulses.

> To get an idea for ambient photon fluxes, calculate the number of photons emitted per nanosecond from the visible spectral region of an incandescent 100-watt bulb.

Answers to
Exercises

Answers to Exercise 1

1. This is a classic hidden message. The hidden message was supposedly extracted by the cryptographic group known as *Room 39* run by Sir Alfred Ewing. Take the first letter of each word, insert spaces as appropriate, and you will discover it.

2. This is a neat example of the intricate workings of Poe's mind and his formidable vocabulary. Take the first letter of the first line, the second letter of the second line, the third letter of the third line, and so on, insert spaces as appropriate, and you will find the name

   SARAH ANNA LEWIS

   almost in answer to the challenge in the last line of the sonnet:

   " . . . *the dear names that lie concealed within 't.*"

3. This is the famous letter delivered to John Trevanion. Aside from the dated English, the structure is disturbing in its punctuation. If we take the third letter after each punctuation event, we uncover the message that saved the Cavalier's life:

   PANEL AT EAST END OF CHAPEL SLIDES

4. There would be 27! possible keying variables or more than 10^{28}.

5. We know from the definition of the factorial that 200! is the product of the integers, $1, 2, 3, \ldots, 199, 200$. The number of contiguous ending zeros in this product is governed by the number of factors of 10 that are generated in the large multiplication. To produce a 10 requires a 2 and a 5. There are many 2's but many fewer 5's. Therefore, if we count the number of 5's in the product, we also count the number of contiguous zeros with which 200! ends. To do this, first note that there is a factor of 5 every 5th number and therefore 40 zeros generated from this alone. However, every 25th number contains 5-squared and so every 25th number adds an additional zero to the product. There are 8 such occurrences of this in 200!, so we are up to 48 zeros. Further, every 125th number contains 5-cubed, so we must count an additional zero every 125 numbers. This adds one more zero to our sum so we can say that 200! ends with 49 zeros.

   We can be more analytical by using the operation $\lfloor x \rfloor$ known as the *floor function*. The floor function yields the greatest integer less than or equal to its argument. For example: $\lfloor 2 \rfloor = 2, \lfloor 5.9 \rfloor = 5, \lfloor -3.2 \rfloor = -4$. The number of zeros that N! ends with is then easily seen to be

$$\left\lfloor \frac{N}{5} \right\rfloor + \left\lfloor \frac{N}{5^2} \right\rfloor + \ldots = \sum_{i=1}^{\infty} \left\lfloor \frac{N}{5^i} \right\rfloor$$

6. Consider the first ten numbers in the product for 250!:

$$1 \times 2 \times 3 \times 4 \times 5 \times 6 \times 7 \times 8 \times 9 \times 10$$

There is one 9, but we also have another 9 developed from the product of the 3 and the factor 3 within 6(=2×3). This case requires a slightly deeper thought than did problem 5. The difference is that 9 is not a prime number. The way to count the number of 9's is to count the number of 3's and then take the floor function of half the number of 3's that we find. We take half the number of 3's, because 9 is 3 squared, and we use the floor function because an odd number of 3's will "leave a single 3 left over" as we combine the 3's into pairs. Thus, the number of 9's in 250! is easily determined as follows: First we find the number of 3's in the product 250!.

$$\left\lfloor \frac{250}{3} \right\rfloor + \left\lfloor \frac{250}{9} \right\rfloor + \left\lfloor \frac{250}{27} \right\rfloor + \left\lfloor \frac{250}{81} \right\rfloor + \left\lfloor \frac{250}{243} \right\rfloor = 123$$

Then we find that $n = \left\lfloor \frac{123}{2} \right\rfloor = 61$.

Answers to Exercise 2

1. This is a short sentence (31 letters), and yet each letter makes at least one appearance.

2. There are no *E*s, which is a remarkable tour de force by Wright.

3. The following are plots of the frequency count data identified as a small sample and the data of Table 2-1 identified as a large sample. Note that the statistics of even a small sample of English text is remarkably close to the statistics of a much larger sample.

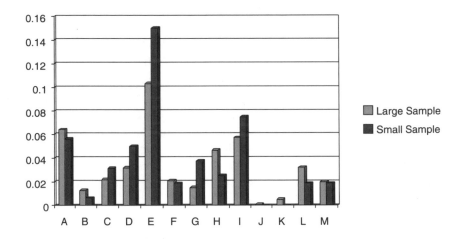

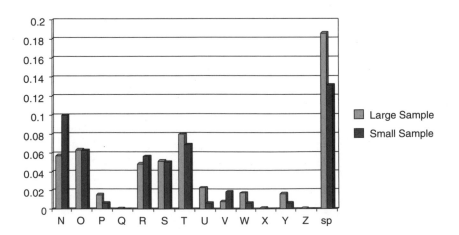

4. The two characters

are the most common. There is a good chance that they correspond to space and *E*, or vice versa. From this assumption, you can proceed to posit other hypotheses and quickly solve the cryptogram to read as follows:

"THERE ARE NO ATHEISTS IN THE FOXHOLES"

Incidentally, this famous quote comes to us from a field sermon given in 1942 on Bataan by Father William Cummings of the Chaplain Corps. Also, this particular cipher has a venerable standing. Both the cipher and its variants are known as the *Pig Pen Cipher*. In the form used in this problem set, it is performed by first writing the plaintext elements in a nine-region space similar to the tic-tac-toe template.

ABC	DEF	GHI
JKL	MNO	PQR
STU	VWX	YZ_

A three-letter group is represented by the boundary of the region in which it lies and the particular member of the three-letter group is denoted by either no dots if the letter is the first member of the group, one dot if it is the second member, and two dots if it is the third.

5. Spectrogram A shows the sound of whispered speech. Notice that the formants of the voiced portions are still visible, although they are spectrally broadened and bathed in noise by the whisper.

Spectrogram B shows the sound of random noise. Notice the broadband nature of the signal and the concomitant lack of structure.

Spectrogram C shows the sound of crumpling paper. Notice that it appears to be made up of a dense series of broadband events interleaved with quiet periods. This is, of course, the snapping and crinkling of the paper.

Spectrogram D shows the sound of a telephone bell. Note that the bell evidently has two gongs whose center frequencies are slightly different and that a low-frequency noise is present, which is probably related to the clapper dynamics.

Spectrogram E shows the sound of piano music. Notice that the events have sharply defined simultaneous beginnings of sets of tones.

6. The source appears, and this is admittedly a very small sample, to never produce two or more contiguous 1's. A 1 is always followed by a 0 and can be likened to the letter Q in the English language, which is invariably followed by a U. Thus, if the source produces a 1, we know that the next bit will be a 0; therefore, the following 0 appears to carry or convey no information. We can strongly hypothesize that the first 60 bits out of the source represent only 40 bits of information.

Answers to Exercise 3

1. According to the problem, the analog signal, *s*, is sampled within a quantizer step region and we may assume that the value of the analog signal is uniformly distributed over the region. Let us assume that the midrange value of the quantizer step region is at s_0 volts. The quantizer outputs this value for the entire step region. Thus, we have the probability density function for *s*, as shown in Figure 3-6.

   The error then has the probability density function shown in Figure 3-7.

   The mean value of *e* is then

$$\bar{e} = \int_{-\Delta/2}^{\Delta/2} \frac{e}{\Delta}\, de = 0$$

and the average power is then

$$\overline{e^2} = \int_{-\Delta/2}^{\Delta/2} \frac{e^2}{\Delta}\, de = \frac{\Delta^2}{12}$$

Figure 3-6
The probability
density function for *S*

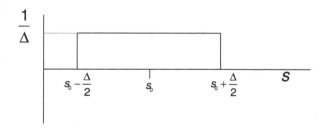

Figure 3-7
The probability
density function for
the error voltage *e*

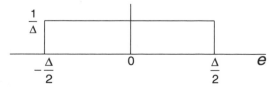

2. If B is unitarily increased, the number of levels is doubled. Therefore, we may consider

$$\Delta \leftarrow \frac{\Delta}{2}$$

In this case,

$$\overline{e^2} \leftarrow \frac{1}{4}\frac{\Delta^2}{12}$$

and the noise power decreases by 6 dB. This is why many people claim that the signal-to-(quantizing) noise ratio is improved by 6 dBs each time the number of the quantizing levels doubles.

Answers to Exercise 4

1. Subtraction is *not* associative. For an operation ∘ to be associative, the following condition must hold:

$$e_i \circ (e_j \circ e_k) = (e_i \circ e_j) \circ e_k$$

where e_i, e_j, and e_k are any three elements. If subtraction were associative, then

$$e_i - (e_j - e_k) = (e_i - e_j) - e_k$$

which is not true.

2. The multiplication of square matrices is *not* commutative. For an operation ∘ to be commutative, $e_i \circ e_j = e_j \circ e_i$ for any two elements (e_i and e_j). For two square matrices,

$$A = \begin{pmatrix} a & b \\ c & d \end{pmatrix} \text{ and } B = \begin{pmatrix} w & x \\ y & z \end{pmatrix}$$

$$AB = \begin{pmatrix} aw + by & ax + bz \\ cw + dy & cx + dz \end{pmatrix} \text{ and } BA = \begin{pmatrix} aw + cx & bw + dx \\ ay + cz & by + dz \end{pmatrix}$$

3. The operator f, when used three times as described, exchanges the contents of the memory locations A and B.

4. If $p(A)$ represents the probability of event A, and $p(B|A)$ represents the probability of event B given the occurrence of event A, then

$$p(A|B)p(B) = p(B|A)p(A)$$

Bayes' Theorem results from writing the previous equation in the following form

$$p(A|B) = \frac{P(B|A)p(A)}{p(B)}$$

If event A represents that both children are girls and event B represents that at least one child is a girl, then

$$p(A) = \frac{1}{4}$$

$$p(B) = \frac{3}{4}$$ (This is unity minus the probability that both children are *a priori* boys.)

$$p(B|A) = 1$$

and we have $p(A|B) = \dfrac{1 \times \dfrac{1}{4}}{\dfrac{3}{4}} = \dfrac{1}{3}$

Note that you must pay particularly close attention to the formulation of a probability question. Consider the following question: A couple has two children. The older child is a girl. What is the probability that both of the children are girls? If this was the question, then the answer would have been one-half.

5. One way to use a biased coin is to flip it twice. If it comes down as heads-heads or tails-tails, then flip it twice again, and so on until you obtain either heads-tails or tails-heads. Assign a zero to heads-tails and a one to tails-heads. Zeros and ones are equally probable as the probability of heads-tails is $p(1-p)$ and the probability of tails-heads is $(1-p)p$.

6. As noted in the discussion following Equation 4.7, if one of the sources is a balanced binary Bernoulli source, the output is indistinguishable from a balanced binary Bernoulli source. Thus, when the five sources of the problem are combined, the presence of the source whose p is 0.5 turns the resultant p into a balanced binary Bernoulli source and $p = 0.5000$.

Answers to Exercise 5

1. The plaintexts are as follows:

CLOSE THE DOORS

and

SHUT THE WINDOW

2. Two plaintext characters are in the same position if the mod-2 bit-by-bit sum of the ciphertexts in depth is 00000. This occurs four times.

3. The average number of bits per symbol, $\bar{l}$, will be

$$\bar{l} = \sum_i p_i \times b_i$$

where p_i is the probability of symbol i, and b_i is the length of the Huffman code word for symbol i, $\bar{l} = 4.1195$, which is a reduction of almost 18 percent over the coding used in Figure 3-1 of Module 3.

If the plaintext were first encoded by a Huffman code, the problem of solving for the plaintexts in a depth situation would be far more complicated because we wouldn't know where the boundaries of the plaintext symbols are as the Huffman code is a variable-length code.

4. There were 35 bits of ciphertext sent and therefore 7 characters of plaintext. There are 3 possible candidates: FOUR_PM, FIVE_PM, and NINE_PM, each with an a priori probability of one-third. This is an example of a traffic-flow security problem. The messages may be properly encrypted, but the security is mitigated by the overall system procedures.

Answers to Exercise 6

1. First of all, you have experienced a mathematical result and *not* a card trick. The reason that you were unable to affect the density of the suits in successive four-tuples relates to the structure of the two partial decks that you shuffled. Remember that you built the second deck by taking cards off the top of the original deck one at a time. The second pile had the suit sequence reversed. For this reason, no matter how your riffle shuffle proceeded, you always end up with one of each suit in every group of four cards. To see this graphically, consider the progression of the following shuffle shown in Figure 6-3.

   No matter how you select the four cards, you must encompass one of each suit.

2. The FLAG is 01111110. For this sequence, $e_1 = e_8 = 1$ and all other $\{e_i\}$ are zero. Using Equation 6.7, we find that the mean waiting time for the appearance of the FLAG is $d(2) = 258$ bits.[5]

Figure 6-3
An example of the partial results of a riffle shuffle

---

[5]There is actually a slight approximation here as a false flag could be produced by the 7-bit sequence 0111111 occurring immediately before the final flag.

Answer to Exercise 7

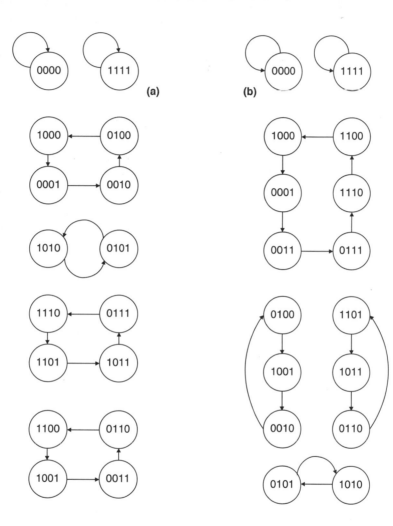

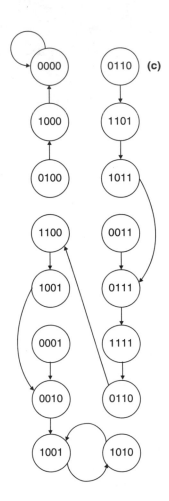

(c)

(d)

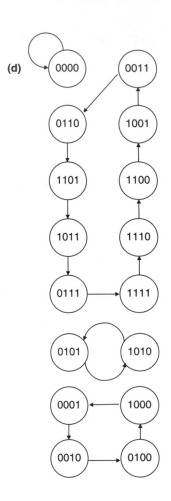

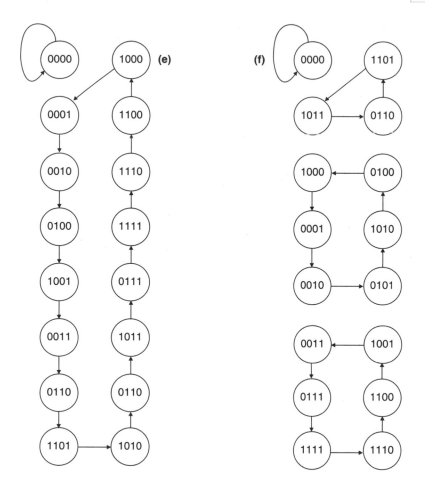

What did you find for Machine (a)? Its 16 states seem to be partitioned into six groupings. In each of these groupings, we note that each state has a single predecessor; that is, exactly one state directly precedes it. Such a grouping of states is called a *cycle*. Therefore, Machine (a) exhibits six cycles: two of length 1, one of length 2, and three of length 4. Machine (a) generates what are called *necklace cycles*, which are cycles that are simply cyclic end-around shifts of 4 bits. In each cycle, each state has the same number of ones.

Machine (b)'s behavior is a bit more complex. The exclusive-or tree feeds back a function of the register's contents to help form the next state. If we number the stages with the convention shown in Figure 7-7, we can write

Figure 7-7
Stage numbering
convention

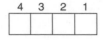

a set of equations that describe the state transitions. We let x_i^t represent the content of stage i at time t. Then we have

$$
\begin{aligned}
x_1^{t+1} &= x_1^t \oplus x_3^t \oplus x_4^t \\
x_2^{t+1} &= x_1^t \\
x_3^{t+1} &= x_2^t \\
x_4^{t+1} &= x_3^t
\end{aligned}
\tag{7.1}
$$

Note that Machine (b) exhibits six cycles.

Machine (c)'s behavior is very different from the machines just examined. Machine (c) has states that do not form cycles. A number of the states have more than one predecessor and, consequently, a number of states have no predecessor. Let's write the state transition equations for Machine (c) as we did for Machine (b). The two types of logic gates used in Machine (c) are the exclusive-or gate and the *and* gate, which is simple multiplication.

$$
\begin{aligned}
x_1^{t+1} &= x_4^t x_3^t \oplus x_2^t \\
x_2^{t+1} &= x_1^t \\
x_3^{t+1} &= x_2^t \\
x_4^{t+1} &= x_3^t
\end{aligned}
\tag{7.2}
$$

From its schematic in the exercise, Machine (d) appears to be very similar to Machine (c), but there is a crucial, very important difference. Machine (d) exhibits cycles; Machine (c) does not. Why?

The reason can be found in the equations for state transition. For Machine (d), they are

$$
\begin{aligned}
x_1^{t+1} &= x_4^t \oplus x_3^t x_2^t \\
x_2^{t+1} &= x_1^t \\
x_3^{t+1} &= x_2^t \\
x_4^{t+1} &= x_3^t
\end{aligned}
\tag{7.3}
$$

Can you summarize the difference between Equations 7.2 and 7.3? It has to do with the manner in which the content of the last stage, stage 4, enters into the feedback. Consider that by definition a cycle has states that have unique predecessors. This implies that a machine with cycles can theoretically be run forwards or backwards in time. Machine (d) has cycles because we can write the equations to determine the previous state.

$$
\begin{aligned}
x_1^t &= x_2^{t+1} \\
x_2^t &= x_3^{t+1} \\
x_3^t &= x_4^{t+1} \\
x_4^t &= x_1^{t+1} \oplus x_4^{t+1} x_3^{t+1}
\end{aligned}
\tag{7.4}
$$

If we try to do the same with Machine (c), we arrive at a quandary because we cannot uniquely solve for x_4^t. If $x_3^t x_4^t = 0$, then x_4^t is unknowable. The lesson is that a machine exhibits cycles if and only if x_4^t enters the feedback function for x_1^{t+1} in a linear fashion.

Now finally let's examine Machines (e) and (f). Machine (f) exhibits four cycles and Machine (e) exhibits two cycles. Both machines exhibit a one-step cycle that comprises the all-zero state. First of all, why must the all-zero state constitute a cycle? The answer is obvious. The feedback function is composed of one or more exclusive-or gates. These gates are, as we recall, addition mod-2; therefore, if their inputs are all zeros, then their outputs are also zeros. Thus, with mod-2 feedback, the all-zero state must be a one-step cycle. However, Machine (e) has a property that is truly remarkable. It links all the other 15 states together in a single cycle. This is termed an *m-sequence generator*, which we study in the next module.

Answers to Exercise 8

1. The machine in Figure 8-4 has eight possible settings for the three switches. The output sequences corresponding to the settings are summarized in Table 8-4.

   The machine sketched in Figure 8-4 is capable of producing the all-zero sequence and all seven phases of the m-sequence according to $x^3 + x + 1$.

2. The key to solving this problem is to recognize that the machine we are dealing with is a linear machine. As such, we may use superposition as we did in Equation 7.7 of Module 7. We could, for example, combine the two initial register settings: 0100111010 and 0111010010. The result would be another initial register setting, and the register contents that would follow this new register setting after 500 steps would be the bit-by-bit mod-2 sum of the register contents that followed the two initial register settings that were combined. This is shown in Figure 8-6.

   The following question arises: Is there a combination of the four initial register settings that will yield the initial setting of 1011000101? The

Table 8-4

The eight output sequences of the machine in Figure 8-4

S_1	S_2	S_3	Output Sequence
Open	Open	Open	0000000
Open	Open	Closed	1001110
Open	Closed	Open	0011101
Open	Closed	Closed	1010011
Closed	Open	Open	0111010
Closed	Open	Closed	1110100
Closed	Closed	Open	0100111
Closed	Closed	Closed	1101001

Figure 8-6

Combining two initial register settings

old initial register setting	0100111010	after 500 steps	1001000001
old initial register setting	0111010010	after 500 steps	1101000111
new initial register setting	0011101000	after 500 steps	0100000110

Figure 8-7
Using superposition
in a linear machine
structure

0100111010 1001000001
0111010010 1101000111
1000101101 1011011011
1011000101 **1111011101**

answer is yes—if we combine the initial settings 1, 2, and 4, we will obtain 1011000101. The state that must follow 1011000101 after 500 steps is therefore the bit-by-bit mod-2 sums of the states that followed the initial settings 1, 2, and 4 after 500 steps. This is shown in Figure 8-7. The answer is in bold.

Notice that specifying the m-sequence generator polynomial, $x^{10} + x^3 + 1$, was simply gratuitous. So long as we know that the machine is linear, it doesn't matter what its particular state transition behavior is.

3. An efficient way to open the lock, if we know the value of n, is to simply feed an m-sequence of length $2^n - 1$ into the input register. If the m-sequence is started at a randomly chosen phase, we can expect to unlock the lock in approximately 2^{n-1} steps.

Answers to Exercise 9

1. By bit-by-bit mod-2 adding the plaintext bits to the ciphertext bits we get the following keytext bits:

$$\begin{aligned}
&\text{CT:}\ 1\ 1\ 1\ 1\ 1\ 1\ 0\ 1\ 0\ 1\ 0\ 0\ 0\ 0\ 1 \\
&\underline{\text{PT:}\ 0\ 0\ 0\ 0\ 0\ 0\ 1\ 1\ 0\ 1\ 0\ 0\ 0\ 1\ 1} \\
&\text{KT:}\ 1\ 1\ 1\ 1\ 1\ 1\ 0\ 0\ 0\ 0\ 0\ 0\ 0\ 1\ 0
\end{aligned}$$

Note that the keytext has a run of six zeros. A run of six zeros can be present only if $n \geq 7$.

2. We form the candidate keytexts in Figure 9-5 for the five pairs of plaintext letters in the same manner as in the previous problem.

If we have correctly guessed the plaintext, then the keytext will obey the recursion rule of the primitive polynomial $x^5 + x^2 + 1$ and the three pointers (x, y, and z) will be able to be slid along the putative keytext and always satisfy the relation $z = x \oplus y$. We see that this relation is not satisfied for SO, ME, DR, or AN, but it is satisfied for LA, the viable candidate.

Figure 9-5

The five possible plaintexts

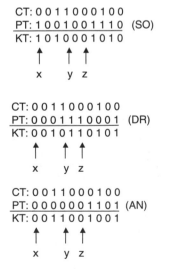
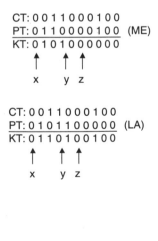

3. Again, we solve for the keytext first:

$$\text{CT: } 0\ 0\ 1\ 1\ 0\ 0\ 0\ 0\ 1\ 1$$
$$\text{PT: } 1\ 0\ 0\ 1\ 1\ 0\ 0\ 1\ 1\ 1$$
$$\text{KT: } 1\ 0\ 1\ 0\ 1\ 0\ 0\ 1\ 0\ 0$$

As in the previous problem, we can think of pointers for the recursion sliding along the keytext. In this case, the pointers are the coefficients of the polynomial: $c_5 c_4 c_3 c_2 c_1$. We start by lining c_5 under the bit farthest on the left: a 1. We multiply the coefficients by the value of the bit that they remain under, add the terms of the five individual multiplications, and set the result equal to the next bit in the keytext. Thus, we develop the following set of equations:[3]

$$c_5 + c_3 + c_1 = 0 \qquad\qquad (9.1)$$

$$c_4 + c_2 = 0 \qquad\qquad (9.2)$$

$$c_5 + c_3 = 1 \qquad\qquad (9.3)$$

$$c_4 + c_1 = 0 \qquad\qquad (9.4)$$

$$c_5 + c_2 = 0 \qquad\qquad (9.5)$$

By inspection, we see that by combining Equations 9.1 and 9.3, we find that $c_1 = 1$; therefore, by using Equation 9.4, $c_4 = 1$. By using Equation 9.2, $c_2 = 1$. By using Equation 9.5, $c_5 = 1$. And, finally, by combining Equations 9.1 or 9.3, $c_3 = 0$.

Another more elegant way to solve the problem is to cast it as a linear system and cast Equations 9.1 through 9.5 as a matrix B as follows:

$$B = \begin{bmatrix} 1 & 0 & 1 & 0 & 1 \\ 0 & 1 & 0 & 1 & 0 \\ 1 & 0 & 1 & 0 & 0 \\ 0 & 1 & 0 & 0 & 1 \\ 1 & 0 & 0 & 1 & 0 \end{bmatrix} \qquad\qquad (9.6)$$

---

[3]If we wanted, we could have saved some time and assumed that $c_5 = 1$ because we were told we were dealing with a five-stage shift register with linear feedback.

We then form the linear (mod-2) set of equations:

$$(c_5\, c_4\, c_3\, c_2\, c_1)B = (0\ 0\ 1\ 0\ 0) \tag{9.7}$$

We then invert the matrix B:

$$B^{-1} = \begin{bmatrix} 1 & 1 & 1 & 1 & 1 \\ 1 & 0 & 1 & 1 & 0 \\ 1 & 1 & 0 & 1 & 1 \\ 1 & 1 & 1 & 1 & 0 \\ 1 & 0 & 1 & 0 & 0 \end{bmatrix} \tag{9.8}$$

Finally, we right-multiply both sides of Equation 9.7 by B^{-1} and find that

$$(c_5\, c_4\, c_3\, c_2\, c_1) = (0\ \ 0\ \ 1\ \ 0\ \ 0)B^{-1}$$

$$= (1\ \ 1\ \ 0\ \ 1\ \ 1) \tag{9.9}$$

Answers to Exercise 10

1. The key to answering this question comes from Module 7 and the crucial difference between Machines (c) and (d). Machine (d) could be run backwards because the content of its last stage entered the feedback function in a linear fashion. The same principle is used here. Note that a round has a left-hand and right-hand side. The right-hand side is operated on by the nonlinear function $f(\cdot)$ to produce pseudorandom bits that are bit-by-bit mod-2 added to the left-hand side, which then becomes the new (lower) right-hand side. The old right-hand side, however, becomes the new (lower) left-hand side. To go backwards (up the encryption process of Figure 10-4), we simply use the lower left-hand side to generate the bits needed to be added to the lower right-hand side to recreate the new (upper) left-hand side. Thus, to back up the DES (that is, to compute CIPH_k^{-1}), we can, at each stage, find the output of the nonlinear function $f(\cdot)$ and bit-by-bit mod-2 add it to the lower right-hand side to recover the upper left-hand side.

2. The answer to this problem is very simple: backwards compatibility. Many DES devices are in place as fixed plant equipment. If you have a triple DES device and you want to make it compatible with a single DES device, just set $K_2 = K_3$ and the device will act as a single DES device keyed by the keying variable K_1.

Answers to Exercise 11

1. We assume that all members of the net use the same secret keying variable. We want to avoid the situation where a b-bit input is reused in the input block of the block cipher function as this would cause a depth.

   A straightforward way to do this is to first calculate k as

   $$k = \lceil log_2 M \rceil \qquad (11.4)$$

   where $\lceil x \rceil$ is the *ceiling function* of argument x. This function yields the smallest integer that is equal to or greater than x. We then simply assign the first k of the block cipher function b-input bits in a unique way to each user. For example, the first user would be assigned 00 . . . 00, the second has 00 . . . 01, the third has 00 . . . 10, and so on. These assignments would remain invariant over the *cryptoperiod*, which is the duration of time that a particular secret keying variable is in use. The remainder of the $b-k$ input bits would be from an incrementing counter or other nonrepeating finite state sequential machine.

2. This problem is related to the celebrated *birthday problem*. This is an important problem because its result teaches us something that is widely encountered in one form or another and whose solution is counterintuitive.

   The birthday problem gets its name from the problem formation. It is usually presented something like this:

   In a group of n individuals chosen at random, what is the probability that no two people have the same birthday?

   Some mild assumptions are usually made, such as an absence of twins in the group. Leap year is generally not dealt with, so that leaves 365 dates over which to consider the birthdays distributed.

   The solution proceeds as a simple sequential probability formulation. Let P_ϕ represent the probability that no two birthdays of the n individuals fall on the same day. Then we simply write

   $$P_\phi = \frac{365}{365} \times \frac{364}{365} \times \frac{363}{365} \cdots \frac{365 - n + 1}{365} \qquad (11.5)$$

Figure 11-3
P_ϕ versus n

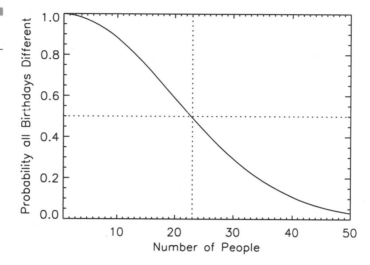

Graphing P_ϕ versus n we come to a remarkable conclusion. Examine the graph in Figure 11-3.

Note that n (the number of randomly chosen individuals) needs to be only 23 before P_ϕ drops below one-half. Twenty-three distinct birthdays is less than 10 percent of the possible birthdays in a year.

The same effect causes the capacity of some random broadcast protocols to be very low. The slotted ALOHA protocol, for example, has a maximum spectrum utilization of about 37 percent. In this protocol, equal duration time slots are used by transmitters transmitting at random. When two or more transmitters transmit during the same time slot, a *collision* occurs. When the system detects a collision, it requires all the colliding transmissions to be resent (with individual random backoffs in time to prevent immediate recollisions). The slotted ALOHA protocol's advantage is that it requires very little overhead to implement and the protocol may be an excellent candidate for a transmitter population that varies unpredictably with time. The downside is that of significant spectrum is underutilization.

Let's move on to the answer for problem 2. We solve the birthday problem for the case where we have $B = 2^b$ possible b-bit input block settings of the block cipher function. We assume that the region of interest enables the assumption $B \gg n \gg 1$ and we use P_ϕ again to represent the probability that none of the n b-bit input contents will be repeated in n encryptions.

Proceeding exactly as before,

$$P_\phi = \frac{B}{B} \times \frac{B-1}{B} \times \frac{B-2}{B} \dots \frac{B-n+1}{B}$$

$$= \frac{1}{B^n} \frac{B!}{(B-n)!}$$

(11.6)

Now we delve way back to Module 1, "First Considerations," and call up Stirling's approximation to the factorial. We have

$$\frac{B!}{(B-n)!} \cong \sqrt{\frac{2\pi B}{2\pi(B-n)}} \frac{B^B}{e^B} \frac{e^{B-n}}{(B-n)^{B-n}}$$

(11.7)

$$\cong e^{-n} \frac{B^B}{(B-n)^{B-n}}$$

and

$$\frac{B^B}{(B-n)^{B-n}} = \frac{B^n \times B^{B-n}}{(B-n)^{B-n}}$$

(11.8)

$$= B^n \left(\frac{B}{B-n} \right)^{B-n}$$

and

$$\left(\frac{B}{B-n} \right)^{B-n} = \left(\frac{1}{1 - \dfrac{n}{B}} \right)^{B-n}$$

(11.9)

Let

$$y = \left(\frac{B}{B-n} \right)^{B-n} = \left(\frac{1}{1 - \dfrac{n}{B}} \right)^{B-n}$$

(11.10)

then we have

$$\ln y = (B - n)\ln\left(\frac{1}{1 - \dfrac{n}{B}}\right)$$

$$\cong (B - n)\ln\left(1 + \frac{n}{B} + \frac{n^2}{B^2} + \cdots\right) \qquad (11.11)$$

$$\cong (B - n) \times \frac{n}{B}$$

$$= n - \frac{n^2}{B}$$

Finally, collecting terms, we find that

$$P_\phi \cong e^{-n^2/B} \qquad (11.12)$$

Equation 11.12 indicates that there will be little chance of reusing a particular input block for the values associated with modern block cipher functions, which have astronomical values such as $B = 2^{128}$.

Answer to Exercise 12

1. Because you know or have correctly guessed the plaintext underlying the ciphertext, you can change what will be decrypted as plaintext by modifying the ciphertext at the appropriate positions.

   The OFB mode simply adds keytext to the plaintext. The keytext is not a function of the previous ciphertext as is the case with the *cipher feedback* (CFB) mode, which has not yet been described. To change the T in NOT READY to a W, we need to invert those bits that are different between a T and W. In Figure 3-1, we see

   $$
   \begin{array}{rl}
   T & - \ 10011 \\
   W & - \ \underline{10110} \\
   & \ \ \ 00101
   \end{array}
   $$

   Therefore, if we invert the 13th and 15th ciphertext bits, we can change the decrypted message as desired.

Answers to Exercise 13

1. An error in the i-th ciphertext bit causes an error in the i-th decrypted plaintext bit. The first block cipher function's output will be corrupted until the i-th ciphertext bit has progressed through the b-stage input register; therefore, the $i + b$-th decrypted plaintext bit may be in error. The second block cipher function's output will be corrupted until all the corrupted bits have passed through the input register of the first block cipher function. As the $i + b$-th decrypted plaintext bit may be in error, the second block cipher function cannot be sure that it has errorless content in its input register until n-bit times after the $i + b$-th bit; therefore, the largest value, j, for which the plaintext bit, $pt(i + j)$, may be in error is $j = b + n$.

2. The adversary can be successful. To change or spoof the message successfully, your adversary would have to change the 15th, 16th, and 17th ciphertext characters. These characters involve bits 76 through 90. Keytext bits 76 through 90 were produced by the block cipher function operating on ciphertext bits 1 through 64; therefore, changing bits 76 through 90 does not affect any other bits as the block cipher function only needs to perform two cycles. The first cycle uses the IV to produce keytext bits 1 through 64. Ciphertext bits 1 through 64 are then used to produce keytext bits 65 through 128. The message is only 25 characters long and therefore requires only 125 keytext bits.

Answers to Exercise 14

1. $n_1 = 1989$ and $n_2 = 1496$

a	b	d	r
1989	1496	—	—
1989	1496	1	493
1496	493	—	—
1496	493	3	17
493	17	—	—
493	**17**	29	0

So, $(1989, 1496) = 17$.

2. The canonical decomposition of 100 is found by inspection to be

$$100 = 2^2 \times 5^2 \tag{14.6}$$

Now we consider the decomposition of 100 in the system S. The primes in S are found to be

$$\{4, \ 7, \ 10, \ 13, \ 19, \ 22, \ 25, \ 31, \ 34, 37, \ 43, \ldots \}$$

Incidentally, these primes are called the *Hilbert primes*. We see that 100 can be decomposed into these primes in two distinct ways:

$$100 \ = 4 \times 25 \tag{14.7}$$

and

$$100 \ = 10^2 \tag{14.8}$$

Answers to Exercise 15

1. $3x \equiv 7(12)$ asks for an x such that 12 exactly divides $3x - 7$. But 12 is a composite number and one of its factors is 3. So if 12 were to exactly divide $3x - 7$, then 3 must also exactly divide $3x - 7$ and this is impossible.

   We're more fortunate with $6x \equiv 3 \bmod(15)$. All the coefficients and the modulus have the factor 3 in common so we can divide the coefficients and the modulus by 3. We are left with $2x \equiv 1 \bmod(5)$. We can see that $x \equiv 3 \bmod(5)$ is the solution.

2. Forty-seven raised to the 1395th power looks formidable. However, let's take a look at what reduction modulo a modulus means. The modulus is, in a very real sense, a zero. We say this because it is congruent to zero, and it can therefore be added or subtracted from any number without affecting that number's equivalent in the principal complete residue set of the modulus.

   So, what can we do with 47? Well, we can subtract the modulus from it and then have an equivalent problem. Compute

   $$(-1)^{1395} \bmod(48) \tag{15.12}$$

   This is much simpler and by inspection we see that the answer is $(-1) \bmod(48)$. But we're not quite finished. We are asked to express the answer as a member of the principal complete residue set of the modulus 48. The members of that set are

   $$R_{48} = \{0, 1, 2, \ldots, 46, 47\} \tag{15.13}$$

   To get $(-1) \bmod(48)$ into Equation 15.13, all we need to do is add the modulus. Our answer is

   $$(-1)^{1395} \bmod(48) \equiv 47 \bmod(48) \tag{15.14}$$

3. This next problem is quite different. We are asked to raise the number 4 to a large power and perform a modular reduction mod a power of 4—that is, $1024 = 4^5$. As we successively compute 4^i for ever greater powers of i, what happens when i reaches 5? At that point, when we

reduce by the modulus 1024, we will be left with 0 as a remainder. In other words,

$$4^5 \equiv 0 \bmod(1024) \tag{15.15}$$

Once we have a zero, anything we multiply times that zero will still leave zero so

$$4^{3207} \equiv 0 \bmod(1024) \tag{15.16}$$

4. There are many approaches to solving this next problem. Let's take a straightforward plunge. The modulus looks to be close to a power of 2— that is, $2^7 = 128$, and

$$2^7 = 128 \equiv 5 \bmod(123) \tag{15.17}$$

From Equation 15.17, we see that we can write

$$2^{14} \equiv 25 \bmod(123) \tag{15.18}$$

and by combining Equations 15.17 and 15.18, we can write

$$2^{21} = 2^7 \times 2^{14} = 2^{21} \equiv 125 \bmod(123) \equiv 2 \bmod(123) \tag{15.19}$$

From Equation 15.19,

$$2^{42} = (2^{21})^2 \equiv 4 \bmod(123) \tag{15.20}$$

Using Equations 15.18 and 15.20, we find that

$$2^{56} = 2^{14} \times 2^{42} \equiv 100 \bmod(123) \tag{15.21}$$

Finally,

$$2^{57} \equiv 200 \bmod(123)$$

$$\equiv 77 \bmod(123) \tag{15.22}$$

Answer to Exercise 16

1. The inverse of 3 mod(101) is

$$3^{\phi(101)-1} \, \text{mod}(101)$$

$$= 3^{99} \, \text{mod}(101) \tag{16.15}$$

by the Euler-Fermat theorem. This is calculated easily:

i	3^i mod(101)
1	3
2	9
4	81
8	$97 \equiv -4 \, \text{mod}(101)$
16	16
32	54
64	$88 \equiv -13 \, \text{mod}(101)$

Now,

$$3^{99} = 3^{64} \times 3^{32} \times 3^2 \times 3 \tag{16.16}$$

so

$$3^{99} \, \text{mod}(101) \equiv (-13)(54)(9)(3)\text{mod}(101)$$

$$\equiv 34 \, \text{mod}(101) \tag{16.17}$$

Answers to Exercise 17

1. **a.** $x = 3$ or 4

   b. $x = 16$

   c. No solution

2. The first thing we do is express 45 in its unique canonical decomposition into powers of prime numbers:

$$45 = 3^2 \times 5 \tag{17.14}$$

Then we compute $\phi(\cdot)$ as per Equation 17.9:

$$
\begin{aligned}
\phi(45) &= \phi(3^2) \times \phi(5) \\
&= 3 \times 2 \times 4 \\
&= 24
\end{aligned}
\tag{17.15}
$$

3. The divisors of 45 are $\{1, 3, 5, 9, 15, 45\}$. Forming $y = \sum_{d|45} \phi(d)$, we obtain

$$
\begin{aligned}
&\phi(1) + \phi(3) + \phi(5) + \phi(9) + \phi(15) + \phi(45) \\
&= 1 + 2 + 4 + 6 + 8 + 24 \\
&= 45
\end{aligned}
\tag{17.16}
$$

We observe that the sum of the ϕ function for each divisor of 45 yields 45, and if we guessed that

$$y = \sum_{d|y} \phi(d) \tag{17.17}$$

we would be correct. In fact, $\phi(\cdot)$ is the only number theoretic function that satisfies Equation 17.17.

4. This can be done by remembering the binomial theorem:

$$(x + y)^p = \sum_{i=0}^{p} \binom{p}{i} x^{p-i} y^i \qquad (17.18)$$

The key is to remember that

$$\binom{p}{i} = \frac{p!}{i!(p - i)!} \qquad (17.19)$$

and to realize that $\binom{p}{i}$ must have a factor of p for $1 \le i \le p - 1$ because p is prime and i and $p - i$ are both less than p; therefore, none of the numbers in their respective factorials can divide p. Because $\binom{p}{i}$ is divisible by p for $1 \le i \le p - 1$, all of these terms are zero mod(p); therefore,

$$(x + y)^p \equiv (x^p + y^p) \, \text{mod}(p) \qquad (17.20)$$

Now let's look at one of the terms of Equation 17.20:

$$x^p \, \text{mod}(p) \qquad (17.21)$$

Because p is a prime, we know that $\phi(p) = p - 1$; therefore,

$$x^p \, \text{mod} \, (p) \equiv x \times x^{p-1} \, \text{mod} \, (p) \qquad (17.22)$$

$$\equiv x \, \text{mod}(p)$$

Equation 17.22, $x^p \, \text{mod}(p) \equiv x \, \text{mod}(p)$, is true for any x as long as p is prime; therefore, we may rewrite Equation 17.20 as

$$(x + y)^p \equiv (x + y) \, \text{mod}(p) \qquad (17.23)$$

Answer to Exercise 18

1. We prepare the table as per Table 18-1, but forego the counting of multiplications.

N	Y	Z
87	1	17
87	17	17
43	17	17
43	17	87*
43	65*	87
21	65	87
21	65	95*
21	14*	95
10	14	95
10	14	36*
5	14	36
5	14	84*
5	65*	84
2	65	84
2	65	87*
1	65	87
1	65	95*
1	14*	95
0	**14**	95

We find that

$$17^{87} \equiv 14 \bmod(101)$$

Answers to Exercise 19

1. We create a table similar to Table 19-1 but without the s_j column as we are seeking only the inverse.

j	r_j	r_{j+1}	d_{j+1}	r_{j+2}	t_j
0	101	3	33	2	0
1	3	2	1	1	1
2	2	1	2	0	-33
3					34

When r_{j+2} became 0, r_{j+1} was unity; therefore, the inverse to 3 mod(101) exists and is $t_3 \equiv 34$ mod(101).

2. Again we create a table similar to Table 19-1 but without the s_j column as we are seeking only the inverse. Note the use of modular arithmetic to express t_3.

j	r_j	r_{j+1}	d_{j+1}	r_{j+2}	t_j
0	167	41	4	3	0
1	41	3	13	2	1
2	3	2	1	1	-4
3	2	1	2	0	$53 \equiv -114$
4					110

When r_{j+2} became 0, r_{j+1} was unity; therefore, the inverse to 41 mod(167) exists and is $t_4 \equiv 110$ mod(167).

Answer to Exercise 20

1. We could use the binary exponentiation algorithm to work out both of these problems, but sometimes it is instructive and fun to do simple cases by other, more ad hoc approaches.

Let's create a short table of selected powers of 3 mod(101):

i	$3^i \bmod(101)$
1	3
2	9
4	$81 \equiv -20$
8	$97 \equiv -4$
16	16
32	54
64	$88 \equiv -13$

First,

$$3^{70} \bmod(101) = 3^{64} \times 3^4 \times 3^2 \bmod(101) \equiv (-13)(-20)(9)$$
$$\equiv 2340 \bmod(101) \equiv 17 \bmod(101)$$

Next,

$$3^{87} \bmod(101) = 3^{64} \times 3^{16} \times 3^4 \times 3^2 \times 3 \bmod(101) \equiv (-13)(16)(-20)(9)(3)$$
$$\equiv 112320 \bmod(101) \equiv 8 \bmod(101)$$

Answer to Exercise 21

1. We have a mixed-radix counter. The three radii are the $\{c_i\}$, that is, 3,5, and 7. These are the $\{m_i\}$ in Equation 21.1. The set of equations that we want to solve is

$$x \equiv 1 \bmod(3)$$

$$x \equiv 0 \bmod(5)$$

$$x \equiv 6 \bmod(7) \tag{21.6}$$

We want the solution $\bmod(M)$ where

$$M = 3 \times 5 \times 7$$

$$= 105 \tag{21.7}$$

We prepare the auxiliary variables. First we have

$$M_1 = 35$$

$$M_2 = 21$$

$$M_3 = 15 \tag{21.8}$$

and then we solve

$$\rho_1 \times 35 \equiv 1 \bmod(3)$$

$$\rho_2 \times 21 \equiv 1 \bmod(5)$$

$$\rho_3 \times 15 \equiv 1 \bmod(7) \tag{21.9}$$

Equations 21.9 can be easily solved and yield

$$(\rho_1, \rho_2, \rho_3) = (2, 1, 1) \tag{21.10}$$

So, we plug into Equation 21.5 and get

$$x \equiv (2 \times 1 \times 35 + 1 \times 0 \times 21 + 1 \times 6 \times 15)\bmod(105)$$
$$\equiv 55 \bmod(105) \tag{21.11}$$

Answer to Exercise 22

1. If the puzzles were sent out sequentially, then when Party B publicly informed Party A that it had solved puzzle #N, the eavesdropper would simply select the N-th puzzle to solve. The eavesdropper would need to perform only the same amount of work, on average, as Party B. There would be no leverage.

Answer to Exercise 23

1. From Table 20-1 of Module 20, we note that Party A computes

$$3^{70} \bmod(101) \equiv 17 \qquad (23.6)$$

and sends this value to Party B. Party B computes

$$3^{87} \bmod(101) \equiv 8 \qquad (23.7)$$

and sends this value to Party A.

Party B computes

$$17^{87} \bmod(101) \equiv 14 \qquad (23.8)$$

as was done in the exercise for the module on the binary exponentiation algorithm. Let's see if Party A gets the same result by computing $8^{70} \bmod(101)$. We'll use the binary exponentiation algorithm.

N	Y	Z
70	1	8
35	1	8
35	1	64
35	64	64
17	64	64
17	64	56*
17	49*	56
8	49	56
8	49	5*
4	49	5
4	49	25
2	49	25
2	49	19*
1	49	58*
1	14*	58
0	**14**	58

*This indicates that modular reduction is performed as usual.

We see that Party A does indeed have the same result as Party B.

Answer to Exercise 24

1. Again, we have $\left\lceil \sqrt{101} \right\rceil$ and $3^{-1} \bmod(101) = 34$. We seek q such that $3^q \equiv 8 \bmod(101)$. We form the two tables shown in Table 24-2. Note that we did not have to recompute the left table, the table for $3^{11m_1} \bmod(101)$ versus m_1, because this side of the equation remains invariant for the same p and a.

As mentioned before, there is an entry on the left that is the same as the entry on the right, namely 11. The entry on the left corresponds to $m_1 = 7$ and the entry on the right corresponds to $m_2 = 10$. From this information, we can quickly ascertain q from Equation 24.1:

$$q = 7 \times 11 + 10$$
$$= 87 \tag{24.7}$$

Table 24-2

The two tables for performing the split search

m_1	$3^{11m_1} \bmod(101)$	$8 \times 34^{m_2} \bmod(101)$	m_2
0	1	8	0
1	94	70	1
2	49	57	2
3	61	19	3
4	78	40	4
5	60	47	5
6	85	83	6
7	11	95	7
8	24	99	8
9	34	33	9
10	65	11	10

Answer to Exercise 25

1. The first thing to do is calculate $\left\lceil \sqrt{127} \right\rceil$ Secondly, the inverse of the primitive element, x, is easily found to be $x^{-1} \equiv x^6 + 1$. (Consider multiplying both sides of $x^7 + x = 1$ by x^{-1}.)

We now prepare the two tables of Table 25-2.

We note that the entry $x^4 + x^3 + 1$ is common to both tables and corresponds to $m_1 = 9$ and $m_2 = 7$; therefore,

$$q = 12 \times 9 + 7$$
$$= 115 \qquad \qquad (25.16)$$

Table 25-2

Split-search work tables

m_1	x^{12m_1} $\mathbf{mod}(x^7 + x + 1)$	$(x^5 + x^3 + x + 1)(x^6 + 1)^{m_2}$ $\mathbf{mod}(x^7 + x + 1)$	m_2
0	1	$x^5 + x^3 + x + 1$	0
1	$x^6 + x^5$	$x^6 + x^4 + x^2$	1
2	$x^6 + x^5 + x^4 + x^3$	$x^5 + x^3 + x$	2
3	$x^6 + x^5 + x^2 + x$	$x^4 + x^2 + 1$	3
4	$x^5 + x^4 + x^3 + x^2 + x$	$x^6 + x^3 + x + 1$	4
5	$x^6 + x^5 + x^4$	$x^6 + x^5 + x^2$	5
6	$x^6 + x^5 + x^3 + x$	$x^5 + x^4 + x$	6
7	$x^6 + x^5 + x^4 + x^2 + x + 1$	$x^4 + x^3 + 1$	7
8	$x^6 + x^3 + x$	$x^6 + x^3 + x^2 + 1$	8
9	$x^4 + x^3 + 1$	$x^6 + x^5 + x^2 + x + 1$	9
10	$x^6 + x^5 + x^4 + x^3 + x^2 + x + 1$	$x^6 + x^5 + x^4 + x$	10
11	x^5	$x^4 + x$	11

Answers to Exercise 26

1. No. Remember that e and $\phi(n)$ are to be relatively prime so that we can find d. See Equation 26.5.

2. We know that

$$a^x \times b^x = (ab)^x \tag{26.21}$$

and this fact does indeed bear on the RSA authentication scheme that we noted. This is a multiplicative relationship known as the *homomorphic property*. The impact to the RSA authentication technique concerns two messages: M_1 and M_2. If both of these messages are signed, the signatures are

$$M_1^{d_A} \bmod(n_A) \tag{26.22}$$

and

$$M_2^{d_A} \bmod(n_A) \tag{26.23}$$

If someone comes into possession of the two signed messages, he or she can then compute

$$(M_1 M_2)^{d_A} \bmod(n_A) \tag{26.24}$$

and $M_1 M_2$ will appear to be properly signed by Party A. Of course, $M_1 M_2$ may be meaningless. This depends on the information content of messages that Party A may originate. If they are English without compression, then $M_1 M_2$ will most likely be meaningless. On the other hand, if they are randomly derived cryptovariables that Party A signs and sends, then $M_1 M_2$ may well be adjudicated as originated by Party A.

Answer to Exercise 27

1. The first thing we need to do is establish that $(2,4033) = 1$ and $(3,4033) = 1$. Then we proceed to the test for a strong pseudoprime. Starting with $a = 2$, we have

$$4032 = 2^6 \times 63 \qquad (27.6)$$

Checking Equation 27.5a, we determine that

$$2^{63} \equiv 3521 \bmod(4033) \qquad (27.7)$$

Equation 27.5a is not satisfied so we move on to the series of tests specified by Equation 27.5b and find that the second case, $r = 1$, yields

$$2^{63 \times 2} \equiv 4032 \bmod(4033) \qquad (27.8)$$

so Equation 27.5b is satisfied. We conclude that m is either a prime or strong pseudoprime to the base 2.

Moving on to the next base, $a = 3$, we find that

$$3^{63} \equiv 3551 \bmod(4033)$$

$$3^{63 \times 2} \equiv 2443 \bmod(4033)$$

$$3^{63 \times 4} \equiv 3442 \bmod(4033)$$

$$3^{63 \times 8} \equiv 2443 \bmod(4033)$$

$$3^{63 \times 16} \equiv 3442 \bmod(4033)$$

$$3^{63 \times 32} \equiv 2443 \bmod(4033) \qquad (27.9)$$

none of which satisfy Equations 27.5a or 27.5b; therefore, we conclude that 4033 is composite.

Answers to Exercise 28

1. The network keying topology of Figure 28-2 assigns the same keying variable to all links. This makes the key management problem simple, but it opens the network to great exposure for global compromise as there are n places to find the keying variable for the entire net. Let's see what this means quantitatively and then speak further about the security posture.

   First of all, the probability that a compromise will occur at any particular node is assumed to be p; therefore, the probability that no compromise will occur at that node is $1-p$. It is assumed that compromises are independent; therefore, the probability that no compromises will occur within all of the n nodes is

$$(1 - p)^n \tag{28.1}$$

   and the probability that a compromise will occur somewhere in the network, $P_{compromised}$, is

$$P_{compromised} = 1 - (1 - p)^n \tag{28.2}$$

   We plot $P_{compromised}$ in Figure 28-5 for $p = 10^{-6}$ and $p = 10^{-4}$ versus the number of nodes, n, in the range $10 \leq n \leq 1,000$.

   Notice that Figure 28-5 is plotted in a log-log fashion and $P_{compromised}$ is close to linear over the range of n for the representative values of p. This is all very neat and not surprising, but still, look at the numbers

Figure 28-5

$P_{compromised}$ versus n

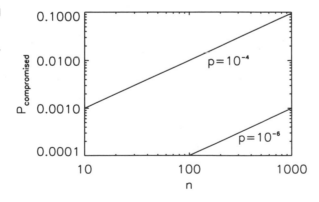

and consider what this means. For the case where p is one in a million, it says that for a net with 1,000 nodes, the chance for compromise is about one in a thousand. Not much? Well, consider that the network supports a large electronic funds transfer system or perhaps a military CINC. Could you really afford to take a one-in-a-thousand chance of compromising all of the funds in transit or all of the sensitive operational traffic? Where does the one-in-a-million value for p come from? Is that the probability that something like a mole has slipped through the screening process? It could also include the probabilities of active hostile physical penetration, loss due to carelessness, or a host of other compromising events. After all that, one in a million might seem a bit low.

The reasons cited previously are why there is such a decided security advantage for using a separate key for each link or even for each session on a link. This is why public key cryptography is so attractive.

2. If you split a 56-bit keying variable between two trustees so that each possesses 28 bits, then you have essentially allowed both trustees to find the entire keying variable and break the ciphertext by trying, on average, only 2^{27} keying variables, which is about 128 million. This is because each trustee starts by knowing half the variable exactly and therefore only has to find the correct 28 bits that the other trustee has. On average, the trustee does this halfway through the total number of possible trials.

Answers to Exercise 29

1. If the first and fourth trustees pool their knowledge, they can reconstitute (k_1, k_2). To do this, they add their first bits together. This yields k_1. To recover k_2, they add the fourth trustee's second bit with the sum of the first trustee's first and second bits.

   If the second and fourth trustee pool their knowledge, they recover (k_1, k_2). This is done by first adding their first bits together. This yields k_2. Next, they add the second trustee's second bit to the sum of the fourth trustee's first and second bits. This yields k_1.

   If the third and fourth trustee pool their knowledge, they recover $(k_1\ k_2)$. First they add the third trustee's second bit to the sum of fourth trustee's first and second bits. This yields k_2. k_1 is then recovered by adding k_2 to the sum of third trustee's first bit and fourth trustee's first bit.

2. First we select shadows one and three and have the simultaneous mixed radix equations:

$$K_0 \equiv 98 \bmod(106)$$

$$K_0 \equiv 53 \bmod(111) \tag{29.17}$$

   Again, using the approach and notation of the Chinese Remainder Theorem, we compute inverses according to

$$\rho_1 M_1 \equiv 1 \bmod(106)$$

$$\rho_3 M_3 \equiv 1 \bmod(111) \tag{29.18}$$

The inverses are

$$\rho_1 = 85$$

$$\rho_3 = 22 \tag{29.19}$$

From these values, we find that

$$K_0 \equiv (85 \times 98 \times 111 + 22 \times 53 \times 106) \bmod(11766)$$

$$\equiv 1052 \tag{29.20}$$

As before, we subtract the largest multiple of the base prime, $p = 101$, from 1052 that we can without making the remainder negative. This multiple is 10 and we are again left with $K = 42$.

Now we select shadows two and three and have the simultaneous mixed radix equations:

$$K_0 \equiv 71 \bmod(109)$$

$$K_0 \equiv 53 \bmod(111) \tag{29.21}$$

Again, using the approach and notation of the Chinese Remainder Theorem, we compute inverses according to

$$\rho_2 M_2 \equiv 1 \bmod(109)$$

$$\rho_3 M_3 \equiv 1 \bmod(111) \tag{29.22}$$

The inverses are

$$\rho_2 = 55$$

$$\rho_3 = 55 \tag{29.23}$$

From these values, we find that

$$K_0 \equiv (55 \times 71 \times 111 + 55 \times 53 \times 109) \bmod(12099)$$

$$\equiv 1052 \tag{29.24}$$

We subtract the largest multiple of the base prime, $p = 101$, from 1052 that we can without making the remainder negative. This multiple is 10 and we are again left with $K = 42$.

Answers to Exercise 30

1. p_Σ is the probability of 1. A 1 will appear in the mod-2 sum of n bits if an odd number of the bits are 1s. Therefore, we can write

$$p_\Sigma = \sum_{i \text{ odd}} \binom{n}{i} p^i (1 - p)^{n-i} \tag{30.8}$$

Expand Equation 30.6 by the binomial theorem

$$((1 - p) - p)^n = (1 - p)^n - \binom{n}{1}(1 - p)^{n-1}p + \binom{n}{2}(1 - p)^{n-2}p^2 - \cdots \tag{30.9}$$

and also expand Equation 30.7

$$((1 - p) + p)^n = (1 - p)^n + \binom{n}{1}(1 - p)^{n-1}p + \binom{n}{2}(1 - p)^{n-2}p^2 + \cdots \tag{30.10}$$

Now, if we subtract the right-hand side of Equation 30.9 from the right-hand side of Equation 30.10, we get the expression

$$2\binom{n}{1}p(1 - p)^{n-1} + 2\binom{n}{3}p^3(1 - p)^{n-3} + 2\binom{n}{5}p^5(1 - p)^{n-5} + \cdots \tag{30.11}$$

However, 30.11 is just twice the right hand side of Equation 30.8 and the left-hand side of Equation 30.10 is 1; therefore, collecting it all together, we have Equation 30.2.

2. If n is even, the p_Σ versus p curves rise from 0 to a maximum of 0.5 and then descends to 0. If n is odd, the p_Σ versus p curves rise from 0 and will be at 1 when $p = 1$. This is easy to grasp intuitively. If n is odd, there are $(n - 1)/2$ pairs of exclusively-ored bits. As p approaches 1, the pairs lean to 0 as both bits in the pair are most likely 1. As a result, the $(n - 1)/2$ pairs tend to produce $(n - 1)/2$ 0s that are added to the one bit of the n bits that is not considered to be paired. This bit tends toward 1. Figure 30-8 shows p_Σ versus p for $n = 2$ and $n = 3$.

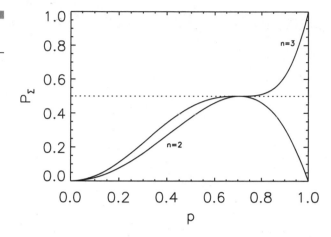

Figure 30-8

p_Σ versus p

Answers to Exercise 31

1. Well, many things could have gone wrong to cause a compromise of a customer's traffic. First of all, an insider in the company could have stolen the keying variable information and conveyed it to the interloper. This is always a possibility and happens often.

   Another possibility is that the cryptographic algorithm has been cryptanalyzed. This is a possibility, of course, and one that should always be considered, but it is not nearly as likely as the following.

   Because the customer can pick the IV, it is possible that the customer was careless, ignorant, or both, and used the same IV for two different messages. If this was the case, operation in the OFB mode places the two different ciphertext messages in depth.

   Finally, the compromise could have been a result of an interloper observing the IV that was used by the customer on the customer's message that the interloper wanted to read (this is the most sinister of these examples). The interloper could then have entered a message for transmission with your service and used the customer's IV. Since the interloper obviously knows the plaintext of his or her own message, he or she can recover the keytext used to encipher it and consequently apply that keytext to recover the plaintext of the customer's message.

   To add insult to injury, the interloper could have also simply copied the customer's ciphertext message, IV and all, and requested that it be transmitted by your service. The ciphertext that the interloper would then observe would be the plaintext of the desired message.

   If the ENCRYPTION4U service had used *cipher feedback* (CFB) as its confidentiality mode, the error extension would not save the day. The interloper could whittle away at the customer's plaintext by sequentially submitting messages to be sent as he or she converged on the plaintext.

2. The parity scheme guards against a particular failure: a break or open circuit in the keying variable insertion path. If this pathway failed in that the bits entered were all zeros or all ones, then the odd parity

would be violated and an error condition would be flagged. Such techniques help to provide an in-depth security environment to keying variable management.

3. The ratio, R, of balanced (equal zeros and ones) k-bit keying variables (k is even) to total k-bit keying variables is

$$R = 2^{-k} \binom{k}{k/2} \qquad (31.4)$$

Using Stirling's approximation, we write Equation 31.4 as

$$R \cong 2^{-k} \frac{\sqrt{2\pi k}\left(\dfrac{k}{e}\right)^k}{\left(\sqrt{\pi k}\left(\dfrac{k}{2e}\right)^{k/2}\right)^2} \qquad (31.5)$$

$$= \sqrt{\frac{2}{\pi k}}$$

4. We know that the coprocessor is used for exponentiation. If the coprocessor uses the binary exponentiation algorithm and a multiplication and modular reduction takes one time unit, then we can observe the number of multiplications required to calculate $\alpha^x \bmod(y)$. This gives away some information about the cryptovariables, thereby reducing the average amount of work that an adversary needs to expend when solving for x by exhausting all possible values.

For example, consider that the coprocessor is calculating $\alpha^x \bmod(101)$ for $1 \le x \le 100$. In Module 18, "The Binary Exponentiation Algorithm," 18.1 gives the number of multiplications required to calculate the exponentiation. For x in the range given, the minimum number of multiplications required is 1. This is the case for $x = 1$. The maximum number of multiplications required is 12 and that is for $x = 95$.

Figure 31-4
Counting the
number of x's
requiring exactly m
multiplications

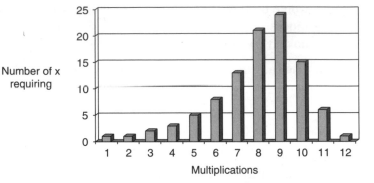

Figure 31-4 shows how many of the numbers in the range $1 \leq x \leq 100$ require exactly m multiplications where $1 \leq m \leq 12$. Note that knowledge of m significantly narrows the number of candidates.

In a high-grade application, x becomes seriously large and a graph corresponding to the graph in Figure 31-4 will look increasingly like a normal distribution. The number of candidates for most values of m will be extremely large, but knowledge of m still provides useful information and disclosure of this should be avoided.

Answer to Exercise 32

1. The graph in Figure 32-8 shows S_P as a function of α for $P = 10$ processors. Note that the speedup drops very rapidly as α moves away from unity. Incidentally,

$$\lim_{P \to \infty} S_P = \frac{1}{1 - \alpha} \tag{32.15}$$

Figure 32-8
S_P versus α for
$P = 10$

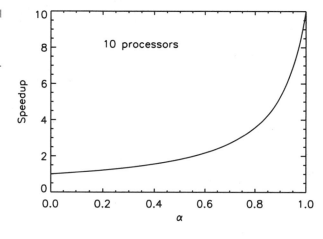

Answer to Exercise 33

1. The binary representation of the input port is 3 = (0,1,1). The desired output port has the binary representation 5 = (1,0,1). The control bits are therefore (1,1,0). We reproduce Figure 33-3 in Figure 33-5 with the control activated according to (1,1,0) and see that the routing algorithm does indeed work as I3 is routed to O5.

Figure 33-5
The omega network
for N = 8

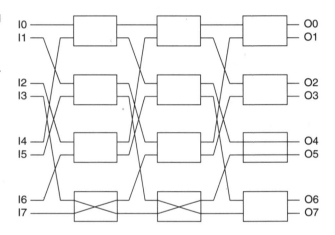

Answer to Exercise 34

1. We have

$$y = \sin(\omega t) + A \sin(\omega t + \theta)$$

$$= \sin(\omega t) + A \cos\theta \sin(\omega t) + A \sin\theta \cos(\omega t) \qquad (34.6)$$

We seek

$$y = B \sin(\omega t + \phi)$$

$$= B \cos\phi \sin(\omega t) + B \sin\phi \cos(\omega t) \qquad (34.7)$$

We equate the quadrature components

$$B \cos\phi = 1 + A \cos\theta$$

$$B \sin\phi = A \sin\theta \qquad (34.8)$$

From Equation 34.8, we take the sum of the squares of the left-hand sides and set it equal to the sum of the squares of the right-hand sides and get

$$B = \sqrt{1 + A^2 + 2A\cos\theta} \qquad (34.9)$$

Then, taking the ratio of both sides, we find that

$$\phi = \tan^{-1} \frac{A\sin\theta}{1 + A\cos\theta} \qquad (34.10)$$

Answers to Exercises 35

1. The problem with long-haul fiber using QC is that you would probably need to install repeater stations and, because a single photon cannot be cloned, repeater stations would have to be intermediate security nodes requiring many overhead duties and security.

   For intermediate-length fiber installations, however, QC may be developed into a valuable network architectural tool. Read the following from a Los Alamos Laboratory release:

   In 1999, researchers at Los Alamos set a record when they sent quantum key through a 31-mile-long optical fiber. While this distance is far enough to create networks connecting closely spaced government offices or local branches of a bank, at greater distances the signal loss in optical fiber increases until the photons are absorbed.

2. The visible spectral region for the human eye is from about 4,000 angstroms to about 7,000 angstroms. An angstrom is 10^{-10} meters. The frequency of a photon, v, of wavelength 4,000 angstroms (the blue end of the color scale) is found by dividing the speed of light by the photon's wavelength:

   $$v_{blue} = \frac{3 \times 10^8 \, m/sec}{4 \times 10^{-7} m} \tag{35.22a}$$

   $$= 0.75 \times 10^{15} \, Hz$$

   The frequency of a photon, v, of wavelength 7,000 angstroms (red end of the color scale) is similarly

   $$v_{red} = \frac{3 \times 10^8 \, m/sec}{7 \times 10^{-7} m} \tag{35.22b}$$

   $$= 0.43 \times 10^{15} \, Hz$$

   The energy of a photon, E, of frequency v is

   $$E = hv \tag{35.23}$$

   where h is Planck's constant and is $h = 6.62 \times 10^{-34} \, joule - seconds$.

Let's assume strictly for convenience that the average photon energy, $\bar{E}$, is midway between the blue end and the red end. Therefore,

$$\bar{E} = 3.91 \times 10^{-19} \, joules \tag{35.24}$$

Now assume an incandescent bulb has an efficiency of 10 percent; that is, assume that 10 percent of the energy expended in the bulb is turned into visible light. In that case, 10 watts are producing visible photons. A watt is a joule-second and so the bulb is producing a photon flux, Φ, of

$$\Phi = \frac{10 \, joules/\sec}{3.91 \times 10^{-19} \, joules} \tag{35.25}$$

$$= 2.6 \times 10^{19} \, photons/second$$

$$= 2.6 \times 10^{10} \, photons/nanosecond$$

GLOSSARY

active attack Manipulation of the communications channel in order to defeat communications security. Active attacks can comprise spoofing, message alteration, and message deletion.

algorithm A series of logical steps, often mathematical in nature, such as an algorithm for finding the square root of a number.

aliasing The error of sampling an analog signal too slowly. Sampling below the Nyquist rate could cause some high frequencies to be aliased to low frequencies.

analog Continuous in time, such as a continuous signal.

balanced Exhibiting ones and zeros with equal probability. A keytext generator should produce a balanced stream of bits.

band splitter A speech privacy device in which the speech spectrum is divided into segments in the frequency domain and the segments are rearranged in a pseudorandom fashion. For example, a band splitter could split the speech spectrum into five different frequency segments and therefore afford $5! = 120$ different permutations of the speech spectrum.

Bayes' Theorem A fundamental law of probability that relates a priori and a posteriori probabilities. Bayes' Theorem relates the probability that event A occurs, given that event B has occurred to the probability that event B occurs, given that event A has occurred.

Bernoulli Having no memory. A fair Roulette wheel exhibits Bernoulli behavior.

block cipher function A cryptographic algorithm that maps b-bit input words into b-bit output words in a one-to-one fashion under control of a cryptographic keying variable. For example, the DES is a 64-bit block cipher function under control of a 56-bit keying variable.

ciphertext Encrypted plaintext. For example, ciphertext should not be understandable by an interceptor.

confidentiality mode A method for using the block cipher function to encrypt and decrypt messages. For example, error extension is an attribute of both the CFB and CBC confidentiality modes.

crib A segment of plaintext that is likely to occur in a message. A cryptanalyst may use a crib as a start in breaking two ciphertexts that are in depth.

cryptanalysis   The set of analytic procedures that are applied in an attempt to gain plaintext from ciphertext without knowing the keying variable.

cryptanalyst   One who attempts to solve or "break" another's protected communications through cryptanalysis. A cryptanalyst might likely be a mathematician.

cryptoalgorithm   An algorithm that specifies a cryptographic principle. For example, a DES cryptoalgorithm is published in FIPS PUB 46.

cryptomachine   Hardware that performs encryption and decryption. Extremely high-speed circuits may require a cryptomachine instead of a software-based approach.

cryptosystem   The equipment or hosted procedure that performs encryption and decryption. A cryptosystem can be entirely realized in software.

cryptovariable   A quantity associated with a particular setting of a cryptosystem. A keying variable and an IV are both cryptovariables.

decryption   The process of converting ciphertext to plaintext. Recovering plaintext from ciphertext is decryption if done by the intended recipient(s), but it is more properly termed cryptanalysis if done by anyone else.

depth   The condition whereby two different plaintext messages are encrypted using the same keytext. A depth condition may compromise both messages.

DES or Data Encryption Standard   A widely used block cipher function.

Doppler shift   The change in frequency of a received signal if the length of the path from the transmitter to the receiver is changing during the signal's propagation from the transmitter to the receiver. For example, due to the Doppler shift, a train's whistle appears to be at a higher frequency when the train is approaching and at a lower frequency when the train is receding.

encoder   A mapping of a set of discrete levels to a set of code words. For example, the speech quantizer levels are converted to binary code words by the encoder.

encryption   The process of converting plaintext to ciphertext. For example, in double encryption, the plaintext of the later encryption is the ciphertext of the earlier encryption.

error extension   The condition wherein a single transmission error creates further errors. For instance, the CFB confidentiality mode exhibits error extension.

Euclidean algorithm   This algorithm finds the greatest common divisor of two numbers. The algorithm can be implemented so that any required divisions are only by 2.

exhaustion   The most basic form of cryptanalysis wherein a cryptanalyst tries all the possible keying variables until plaintext is yielded. Exhaustion is sometimes called brute-force cryptanalysis.

Gaussian   Having a normal probability distribution. Noise samples could be from a Gaussian noise source.

Hamming (7,4) code   Error-correcting code having code words each composed of seven bits and each carrying four information bits. The Hamming (7,4) code is capable of correcting any single bit error.

IV or initialization vector   A cryptovariable, usually not secret, that is used to set the internal state of a cryptosystem. The confidentiality mode specifies how the IV should be managed.

key or keying variable   A secret quantity that sets up or configures a cryptosystem for use.

keytext   A random or pseudorandom stream of bits that appears to be balanced and Bernoulli.

man-in-the-middle attack   An active attack wherein an interloper interposes between two parties who believe they are communicating directly with each other, but in reality they are communicating through the interloper. For example, the Diffie-Hellman system may be subject to a man-in-the-middle attack.

meet-in-the-middle attack   A cryptoanalytic attack wherein one starts with a known, matched plaintext/ciphertext pair and simultaneously works the cryptanalysis from both the plaintext and ciphertext ends so that a viable keying variable match is indicated by a matching of an intermediate result halfway through the encryption and decryption. The meet-in-the-middle attack typically trades between computational time and memory space.

m-sequence   An easily generated pseudorandom sequence. The m-sequence is widely used in communications engineering, particularly in spread spectrum communications and synchronization.

Nyquist sampling criterion   A fundamental rule in signal processing specifying that a signal must be sampled at a rate that is at least twice the frequency of its highest significant spectral content in order to enable perfect reconstruction. For example, the Nyquist sampling rate for voice is generally taken to be 8 kHz.

one-key cryptography   Both parties use the same secret keying variable. One-key cryptography is sometimes called symmetric cryptography.

one-way function   A function of an argument that is relatively easy to compute but relatively difficult to invert. Finite field exponentiation and finite field logarithms are inverses of a one-way function.

passive attack   Monitoring and other nonmanipulative actions.

photon   A basic quantum of light. For example, a photon's energy is equal to the product of Planck's constant and the frequency of the light.

plaintext   Text that is presumed understandable to anyone. Plaintext may be either unencrypted text, speech, or data.

Planck's constant   A basic constant of physics. A photon's momentum is equal to Planck's constant times the ratio of the light's frequency to the speed of light.

pseudorandom   Deterministic but appearing random to someone not discerning the underlying determinism. For example, the advantage of a pseudorandom generator is that its output may be reproduced.

quantizer   A rule for converting samples of a continuous time signal into a finite set of levels. A hard quantizer converts samples into two levels, zeros and ones.

random   Having no underlying determinism, such as a noise source producing a random voltage.

relatively prime   Sharing no common divisor save unity. Neither 77 nor 85 are prime numbers, but they are relatively prime.

sampler (sampling)   Measuring a continuous time signal at specified times. Aliasing is one result of improper sampling.

secret sharing   A technique for encoding a secret number, K, with T different numbers such that t of the T numbers are needed to reconstitute K. The T numbers of a secret sharing system are sometimes called *shadows* and the T personnel holding the shadows are sometimes called *trustees*.

speech inverter   A speech privacy device wherein low frequencies are made high and high frequencies low. A speech inverter can operate to map frequency x into frequency $LO\text{-}x$ where LO is the frequency of a local oscillator that outputs a frequency above the speech frequencies.

spoofing   An attempt to introduce message traffic that appears bona fide in order to mislead or create some nefarious effect.

Stirling's approximation   An easily computed approximation to the factorial. For example, Stirling's approximation involves only a square root and an exponentiation.

subliminal channel   The use of an information channel to convey a secret message along with the traffic normally carried. For example, ending a message with "Thanks" may convey a 0 and ending with "Thank you" may convey a 1.

traffic flow security   A denial of circuit use information to an observer. For example, constantly running a keytext generator may produce traffic flow security.

two-key cryptography   Public key or asymmetric cryptography. In public key cryptography, both parties randomly select secret variables, but each party never knows nor needs to know the other party's secret variable.

two-part code   A cryptosystem widely used in World War I in which one part consisted of plaintext phrases or other plaintext elements and were listed alphabetically with their corresponding ciphertext equivalents. The other part consisted of the ciphertext elements listed lexicographically with their corresponding plaintext equivalents.

INDEX

ABOUT THE AUTHOR

John E. Hershey has more than 30 years of experience in telecom security. The author or co-author of five advanced texts [*Hadamard Matrix Analysis and Synthesis: With Applications to Communications and Signal/Image Processing*; *The Elements of System Design*; *Data Transportation and Protection (Applications of Communications Theory)*; *Perspectives in Spread Spectrum*; and *Doppler Applications in LEO Satellite Communication Systems*], he was elected a Fellow of the IEEE "for contributions to secure communications" and currently works at General Electric's Global Research Center and teaches cryptography at Rensselaer Polytechnic Institute in Troy, New York. He lives in Ballston Lake, New York.